Ergodic Theory of Equivariant Diffeomorphisms: Markov Partitions and Stable Ergodicity

MEMOIRS
of the
American Mathematical Society

Number 803

Ergodic Theory of Equivariant Diffeomorphisms: Markov Partitions and Stable Ergodicity

Michael Field
Matthew Nicol

May 2004 • Volume 169 • Number 803 (end of volume) • ISSN 0065-9266

American Mathematical Society
Providence, Rhode Island

2000 *Mathematics Subject Classification.*
Primary 37A25, 37B10, 37C80, 37D30; Secondary 37C70.

Library of Congress Cataloging-in-Publication Data

Field, Mike.
Ergodic theory of equivariant diffeomorphisms: Markov partitions and stable ergodicity / Michael Field, Matthew Nicol.
p. cm. — (Memoirs of the American Mathematical Society, ISSN 0065-9266 ; no. 803)
"Volume 169, number 803 (end of volume)."
Includes bibliographical references.
ISBN 0-8218-3599-8 (alk. paper)
1. Ergodic theory . 2. Diffeomorphisms. I. Nicol, Matthew. II. Title . III. Series.
QA3 .A57 no. 803
[QA611.5]
510 s—dc22
[515′.48] 2003070916

Memoirs of the American Mathematical Society

This journal is devoted entirely to research in pure and applied mathematics.

Subscription information. The 2004 subscription begins with volume 167 and consists of six mailings, each containing one or more numbers. Subscription prices for 2004 are $583 list, $466 institutional member. A late charge of 10% of the subscription price will be imposed on orders received from nonmembers after January 1 of the subscription year. Subscribers outside the United States and India must pay a postage surcharge of $31; subscribers in India must pay a postage surcharge of $43. Expedited delivery to destinations in North America $35; elsewhere $130. Each number may be ordered separately; *please specify number* when ordering an individual number. For prices and titles of recently released numbers, see the New Publications sections of the *Notices of the American Mathematical Society.*

Back number information. For back issues see the *AMS Catalog of Publications.*

Subscriptions and orders should be addressed to the American Mathematical Society, P. O. Box 845904, Boston, MA 02284-5904, USA. *All orders must be accompanied by payment.* Other correspondence should be addressed to 201 Charles Street, Providence, RI 02904-2294, USA.

Memoirs of the American Mathematical Society is published bimonthly (each volume consisting usually of more than one number) by the American Mathematical Society at 201 Charles Street, Providence, RI 02904-2294, USA. Periodicals postage paid at Providence, RI. Postmaster: Send address changes to Memoirs, American Mathematical Society, 201 Charles Street, Providence, RI 02904-2294, USA.

This publication is indexed in *Science Citation Index*®, *SciSearch*®, *Research Alert*®, *CompuMath Citation Index*®, *Current Contents*®/*Physical, Chemical & Earth Sciences.*
Printed in the United States of America.

∞ The paper used in this book is acid-free and falls within the guidelines established to ensure permanence and durability.
Visit the AMS home page at http://www.ams.org/

10 9 8 7 6 5 4 3 2 1 09 08 07 06 05 04

Contents

Abstract

We obtain stability and structural results for equivariant diffeomorphisms which are hyperbolic transverse to a compact (connected or finite) Lie group action and construct 'Γ-regular' Markov partitions which give symbolic dynamics on the orbit space. We apply these results to the situation where Γ is a compact connected Lie group acting smoothly on M and F is a smooth (at least C^2) Γ-equivariant diffeomorphism of M such that the restriction of F to the Γ- and F-invariant set $\Lambda \subset M$ is partially hyperbolic with center foliation given by Γ-orbits. On the assumption that the Γ-orbits all have dimension equal to that of Γ, we show that there is a naturally defined F- and Γ-invariant measure ν of maximal entropy on Λ (it is not assumed that the action of Γ is free). In this setting we prove a version of the Livšic regularity theorem and extend results of Brin on the structure of the ergodic components of compact group extensions of Anosov diffeomorphisms. We show as our main result that generically (F, Λ, ν) is stably ergodic (openness in the C^2-topology). In the case when Λ is an attractor, we show that Λ is generically a stably SRB attractor within the class of Γ-equivariant diffeomorphisms of M.

Received by the editor February 13, 2003.

1991 *Mathematics Subject Classification.* Primary 37A25, 37B10, 37C80, 37D30 Secondary 37C70.

Key words and phrases. Stable ergodicity, partially hyperbolic, equivariant, attractor, Bernoulli.

MJF supported in part by NSF Grants DMS-1551704 and DMS-0071735.

MJN would like to thank the Leverhulme Trust for the award of a Research Fellowship to work on this project at the University of Houston, the University of Houston for hospitality and the Royal Society and EPSRC for travel support.

Both authors would like to thank Keith Burns, Mark Pollicott, Andrew Török, Charles Walkden, Lai-Sang Young and Marcelo Viana for helpful conversations and communications.

CHAPTER 1

Introduction

Following the work of Grayson, Pugh & Shub [**35**], there has been considerable interest in stable ergodicity for non-hyperbolic systems. Pugh & Shub [**65**] have asked about when it is possible to establish *openness* or *stability* of ergodicity. In particular, they have suggested that if hyperbolicity holds transverse to a center foliation of M ("partial hyperbolicity"), then stable ergodicity may hold generically. Results along these lines for volume preserving diffeomorphisms are presented in [**35, 65, 74**] (see also Brin & Pesin [**13**], for partial hyperbolicity). We refer the reader to the survey article [**15**] by Burns, Pugh, Shub and Wilkinson for a survey of recent results on stable ergodicity as well as technical background.

A natural context for partial hyperbolicity is that of skew (or principal) extensions of hyperbolic systems by a compact Lie group Γ. In 1975, Brin [**12**] proved the genericity of stable ergodicity for compact Lie group extensions of (transitive) Anosov diffeomorphisms. Specifically, Brin showed that there was an open and dense set of *transitive* compact Lie group extensions of an Anosov diffeomorphism [**12**, Theorem 2.2]. Since the base map is Anosov, it follows from [**13**, Corollary 5.3] that every transitive extension is Kolmogorov and, a fortiori, ergodic.

More recently, using somewhat different methods, Adler, Kitchens & Shub [**2**] reproved a variant of Brin's result that applied to circle extensions of Anosov diffeomorphisms of a torus. Specifically, they showed that if $T : \mathbb{K}^n \to \mathbb{K}^n$ is an Anosov diffeomorphism of the torus $\mathbb{K}^n$, then there is an open (C^0-topology) and dense (C^∞-topology) subset $\mathcal{U}$ of $C^\infty(\mathbb{K}^n, \mathbb{K})$ such that if $f \in \mathcal{U}$ then the map $T_f : \mathbb{K} \times \mathbb{K}^n \to \mathbb{K} \times \mathbb{K}^n$, $T(k, x) = (kf(x), Tx)$, is ergodic. Following this result, Parry & Pollicott [**58**] proved the stability and genericity of mixing for Hölder toral extensions over aperiodic (topologically mixing) subshifts of finite type and for toral extensions of mixing hyperbolic systems subject to a simple cohomological condition. Field & Parry [**31**] proved the stability and genericity of ergodicity (and mixing) for a large class of compact Lie group extensions of mixing hyperbolic systems and also showed that stable ergodicity holds generically for all compact connected semisimple Lie group extensions. While the results of Field & Parry apply to smooth extensions if Γ is semisimple or the extension is of a connected hyperbolic set, the results of Parry & Pollicott for toral extensions over subshifts of finite type apply only to Hölder continuous extensions. Indeed, the space of stably ergodic Hölder extensions contains no smooth – even C^1 - extensions! However, recent work of Field, Melbourne & Török [**29**] gives optimal genericity and stability results for smooth compact connected Lie group extensions over completely general basic sets. In another direction, Burns & Wilkinson [**16**] have shown that stably ergodic compact Lie group extensions over a large class of Anosov diffeomorphisms are stably ergodic within the class of volume preserving diffeomorphisms.

From the viewpoint of equivariant dynamics, and symmetry breaking, it is particularly interesting to study systems where the group action is *not* free. Indeed, when Γ is finite, this situation has been extensively studied, especially in the context of symmetry breaking and bifurcation theory [**33**]. When Γ is compact, it is natural to ask about Γ-invariant subsets Λ of a Γ-equivariant dynamical system which are hyperbolic *transverse* to the action of Γ (see Field [**22, 23**]). Hyperbolicity transverse to group orbits forces the dimension of Γ-orbits in Λ to be constant (and typically equal to the dimension of Γ). However, even if all Γ orbits have dimension equal to that of Γ, the action of Γ on Λ may not be free. In particular, Λ may contain singular orbits – orbits with nontrivial isotropy. If we assume that Γ-orbits in Λ have dimension equal to that of Γ, then singular orbits in Λ have finite isotropy group. We remark that if Λ contains a singular orbit then Λ cannot be conjugate to a skew or principal extension.

In this work, we prove a number of foundational results about the ergodic theory of diffeomorphisms equivariant with respect to a compact Lie group Γ. Specifically, we study compact invariant sets that are hyperbolic *transverse* to the group action or *transversally hyperbolic*. Even when Γ is finite (and so hyperbolicity transverse to the group action is equivalent to hyperbolicity) subtle dynamics can occur. For example, if Γ is finite and Λ is a hyperbolic (basic) set for a Γ-equivariant diffeomorphism, we show that Λ/Γ admits a *finite* Markov partition but typically dynamics on Λ/Γ is not expansive. As a well-known example of this phenomenon, we cite the pseudo-Anosov diffeomorphism of the 2-sphere derived by taking the orbit space quotient of the Thom-Anosov diffeomorphism of the 2-torus $\mathbb{K}^2$ by the group $\mathbb{Z}_2$ generated by the map induced on $\mathbb{K}^2$ by minus the identity map of $\mathbb{R}^2$.

Our paper naturally divides into two parts. The results in part one are the responsibility of the first author, those in part two are the work of both authors. In the first part of the paper (Chapters 3 through 6), we derive the structural theory for 'basic' sets of a Γ-equivariant C^1-diffeomorphism. Our main result is to construct Markov partitions on the orbit space of a basic set. This result may be regarded as a tentative first step towards constructing 'Markov partitions' for more general partially hyperbolic sets. The construction of Markov partitions on the orbit space involves some technical difficulties, even in the case when Γ is finite. This is not so surprising since expansiveness fails on the orbit space and expansiveness is typically used to prove the existence of Markov partitions (for hyperbolic sets). In part two of the paper (Chapters 7 through 9) we use the existence of Markov partitions on the orbit space as an important step in the verification of generic stable ergodicity for 'transversally hyperbolic' basic sets – that is, basic sets that are hyperbolic transverse to the group action. More specifically, we use the results on Markov partitions for an absolute continuity argument used in the proof of a Livšic regularity theorem. In turn, Livšic regularity is used as a key ingredient in our proofs of generic stable ergodicity. We also obtain results on the existence of SRB measures on transversally hyperbolic attractors. Throughout, we allow non-free group actions, but require that group orbits have the same dimension. In particular, our results cover systems, such as twisted products, that cannot be realized as skew products or principal extensions.

We now describe the contents of this work in more detail. For the convenience of the reader who may not be familiar with either equivariant dynamics or equivariant geometry, we devote Chapters 2 and 3 to a review of some basic results on Lie groups, smooth group actions and equivariant dynamics that we need in the sequel. Chapter 2 comprises a rapid overview of some of the main fact about Lie groups, group actions and equivariant dynamics. Readers familiar with Lie theory may still wish to skim quickly through this chapter to check on notational conventions. In Chapter 3, section 3.1, we look more closely at the geometry of Γ-manifolds, including the stratifications by (normal) isotropy type. In section 3.2, we review the theory of Γ-equivariant subshifts of finite type, where Γ is a finite group. We conclude the chapter with brief sections detailing the routine extensions of results on the Ruelle transfer operator and equilibrium states to the equivariant setting.

Chapter 4 is devoted to the theory of Markov partitions for basic sets Λ invariant by a finite Lie group. Readers who are mainly interested in the case of compact connected groups Γ can safely skim through section 4.2 and omit section 4.3 on finite groups. (Note, however, that while the main result of section 4.2 on the existence of Markov partitions on the orbit space is not used later, some of the constructions and definitions are used in Chapter 6.) In section 4.2, we construct Γ-*invariant* Markov partitions on Λ such that the corresponding symbolic dynamics is given by a Γ-equivariant subshift of finite type and the associated coding map preserves isotropy type. In particular, no information about symmetry type is lost in the coding. This type of Markov partition induces Markov partitions on closures of orbit strata and so the dynamics on Λ comes with a natural filtration induced from the Γ-action (this is a characteristic result of equivariant dynamics and is modeled after results in [**20**]). This type of Markov partition is special to *finite* group actions. In the remainder of the chapter, we consider the more difficult problem of determining the dynamics on the orbit space Λ/Γ. Our main result is the construction of Markov partitions on Λ that induce Markov partitions on Λ/Γ and that allow us to construct a symbolic dynamics on Λ/Γ (even though dynamics on Λ/Γ is *not* expansive unless the action of Γ on Λ is free).

In Chapter 5, we begin our study of partially hyperbolic basic sets invariant by a compact (non-finite) Lie group. In section 5.1, we define the concept of *transversal hyperbolicity* (hyperbolicity transverse to the Γ-action). Transverse hyperbolicity implies that Λ is partially hyperbolic with center foliation given by the Γ-orbits. In sections 5.2 – 5.4 we establish some basic properties and constructions associated with transverse hyperbolicity, notably the bracket operation and expansiveness (in the sense of Hirsch, Pugh and Shub [**37**]) and stability. In section 5.5 we present some examples based on twisted products. In section 5.6, we give our definition of a basic set for an equivariant diffeomorphism F. We say a compact F-invariant set Λ is a basic set if it Γ-invariant, hyperbolic transitive to the Γ-action, has a 'local product structure', and the induced map on Λ/Γ is transitive. We conclude the chapter by verifying that a version of shadowing holds and that relative periodic orbits are dense.

In Chapter 6, we define our concept of Markov partition for Γ-basic sets invariant by a compact (connected) Lie group of transformations. We call these partitions of Λ 'Γ-regular' Markov partitions. We show that if Λ admits a Γ-regular Markov partition, then there is a symbolic dynamics on the orbit space. We conclude the chapter and Part I by proving the existence of Γ-regular Markov partitions for an arbitrary

Γ-basic set. Although the proof uses some of the results of Chapter 4, on Markov partitions for basic sets invariant by a finite group, the construction is not at all a simple extension of these results. Indeed, unlike what happens for finite groups, the group Γ *never* acts freely on the set of rectangles of a Γ-regular Markov partition. Indeed, each rectangle is a closed Γ-invariant subset of Λ. A further technical complication is that we need to prove a strong enough result so as to be able to verify the absolute continuity results needed in Part 2.

In the second part of the work we describe persistent ergodic properties of diffeomorphisms equivariant with respect to a compact Lie group. We allow non-free group actions, but require that group orbits have the same dimension. In particular, our results cover systems that cannot be realized as skew products or principal extensions. Our basic assumption will be that of transverse hyperbolicity, that is hyperbolicity transverse to the group action.

For the remainder of the introduction, assume that Γ is a compact connected Lie group acting smoothly on the manifold M. Let h denote Haar measure on Γ and F be a Γ-equivariant diffeomorphism of M. We usually assume that F is C^s, $\infty \geq s \geq 2$, although some of our results hold with weaker smoothness assumptions on F. Let $\Lambda \subset M$ be compact and F- and Γ-invariant. Set $F|\Lambda = \Phi$ and let ϕ denote the map induced by Φ on the orbit space Λ/Γ. We assume that Λ is transversally hyperbolic for Φ and note that if transverse hyperbolicity holds for one smooth lift of ϕ, then it holds for all smooth lifts of ϕ.

Using the local product structure on Λ, we prove a version of the Livšic Regularity Theorem (Theorem 8.1.1) [**45, 46**] (for related generalizations see [**59, 52, 53, 63**]). More precisely, we consider the induced map ϕ on Λ/Γ and consider its symbolic dynamical description as a subshift of finite type. We equip this system with the Parry measure (equivalently the measure of maximal entropy) and Haar lift this measure to obtain an Φ-invariant measure ν on Λ which is a measure of maximal entropy. The measure ν will be an invariant measure for all elements of $\mathcal{D}^s_\phi(\Lambda)$. We prove a Livšic regularity theorem relative to this measure. Although we emphasize the case of the Parry measure, our results hold for the Haar lift of any ϕ-invariant equilibrium state defined by a Γ-invariant Hölder continuous potential on Λ. In Proposition 8.2.1 we generalize results of Brin on the structure of ergodic components of compact group extensions of Anosov diffeomorphisms to our setting.

Our genericity results apply to spaces of equivariant mappings covering ϕ. Before we state our main results, we need to give a more precise description of these spaces. Let $r \in (0, s]$ and $\mathcal{D}^r_\phi(\Lambda)$ denote the space of Γ-equivariant lifts of ϕ which extend to C^r Γ-equivariant diffeomorphisms of M. We give $\mathcal{D}^r_\phi(\Lambda)$ the C^r-topology. If $\Psi \in \mathcal{D}^r_\phi(\Lambda)$, then Ψ is ν-measure preserving. Closely related to $\mathcal{D}^r_\phi(\Lambda)$, we define the space $C^r_\Gamma(\Lambda, \Gamma)$ of C^r 'cocycles' on Λ. Thus, $C^r_\Gamma(\Lambda, \Gamma)$ is the space of C^r maps $f : \Lambda \to \Gamma$ such that $f(\gamma x) = \gamma f(x)\gamma^{-1}$, all $x \in \Lambda$, $\gamma \in \Gamma$. We give $C^r_\Gamma(\Lambda, \Gamma)$ the C^r-topology. Elements of $C^r_\Gamma(\Lambda, \Gamma)$ provide the natural generalization of cocycles to situations where Λ is not a product. Given $f \in C^r_\Gamma(\Lambda, \Gamma)$, we define $\Phi_f \in \mathcal{D}^r_\phi(\Lambda)$ by $\Phi_f(x) = f(x)\Phi(x)$. While we do not claim that every element of $\mathcal{D}^r_\phi(\Lambda)$ may be written in the form Φ_f, for some fixed Φ, it is the case that if $\Psi, \tilde{\Psi} \in \mathcal{D}^r_\phi(\Lambda)$ are sufficiently C^0 close, then there exists $f \in C^r_\Gamma(\Lambda, \Gamma)$ such that $\tilde{\Psi} = \Psi_f$.

THEOREM 1.1. *Let $r \in (0, s]$, $s \geq 2$. There exists an open and dense subset $\mathcal{U}_r$ of $C^r_\Gamma(\Lambda, \Gamma)$ such that for all $f \in \mathcal{U}_r$, (Φ_f, Λ, ν) is ergodic (in fact, Bernoulli if ϕ is topologically mixing). If $r \geq 2$, $\mathcal{U}_r$ is C^2-open in $C^r_\Gamma(\Lambda, \Gamma)$.*

Although we stated Theorem 1.1 in terms of the space $C^r_\Gamma(\Lambda, \Gamma)$ of cocycles, the result applies equally to the space $\mathcal{D}^r_\phi(\Lambda)$ of diffeomorphisms covering ϕ. Theorem 1.1 is proved in Chapter 8 (Theorem 9.1.4) using the Livšic Regularity, Theorem 8.1.1, Proposition 8.2.1 and results from [**29**].

In Theorems 9.2.1, 9.3.1 of Chapter 8, we show that Theorem 1.1 can be strengthened to obtain C^0 openness in case Λ is an attractor. In these cases we only require Φ to be C^1 and the proof is similar to Brin's original construction for extensions of Anosov diffeomorphisms [**12**, §2].

If Λ is an attractor, we show the density of stably Sinai-Ruelle-Bowen (SRB) attractors within the class of smooth (at least C^2) Γ-equivariant diffeomorphisms (Theorem 9.4.3). Our proof of this result depends on a result of Ledrappier and Young [**44**] giving absolute continuity of conditional measures on the unstable manifolds with respect to Lebesgue.

We conclude this introduction by pointing out that there is now quite an extensive set of results on ergodicity and absolute continuity properties of measures on attractors when all Liapunov exponents are non-zero. See, for example, Pugh and Shub [**64**] and the recent work of Alves, Bonatti and Viana [**3, 6**]. All of these works, however, make some kind of hyperbolicity assumption on the central directions. In our case, we are considering diffeomorphisms which restrict to isometries on central directions (on account of the presence of the group action). As Shub and Wilkinson [**72**] have recently illustrated, measure (volume) preserving perturbations breaking this structure can lead to a dramatic breakdown of absolute continuity.

CHAPTER 2

Equivariant Geometry and Dynamics

In this chapter we give an overview of the general theory of dynamical systems equivariant by a compact Lie group as well as discuss some necessary foundational material from the theory of compact Lie groups and Γ-manifolds. Experts on Lie theory and group actions need only skim most of this chapter for notational conventions.

2.1. Lie groups, Γ-manifolds and representations

2.1.1. Compact Lie groups. We review a little of the structure theory of compact Lie groups (for more details and proofs we refer to [**14**] or [**32**]).

Suppose first that Γ is connected. If Γ is abelian, then Γ is isomorphic to the n-torus $\mathbb{K}^n$, where $\mathbb{K}$ denotes the group of complex numbers of unit modulus (or SO(2) or S^1) and $n = \dim(\Gamma)$.

We recall that Γ is *semisimple* if Γ has finite center and Γ is *simple* if Γ contains no proper normal subgroups. The classical compact Lie groups $\mathrm{SO}(n)$, (special orthogonal group, $n \geq 3$), $\mathrm{SU}(n)$, (special unitary group, $n \geq 2$), and $\mathrm{Sp}(n)$ (symplectic group, $n \geq 2$), provide examples of compact connected semisimple groups. The quotient of anyone of these groups by its center defines a simple group.

Example 2.1.1. The center of $\mathrm{SO}(m)$ is $\mathbb{Z}_2 = \langle -I \rangle$ if m is even, and is trivial if $m > 1$ is odd. Hence $\mathrm{SO}(2n+1)$ is simple, $n \geq 1$. ♡

The only other compact simple groups are the exceptional simple Lie groups G_2, F_4, E_6, E_7, E_8. If Γ is simple, then the universal cover $\tilde{\Gamma}$ is compact semisimple.

If Γ is any compact connected Lie group, there exist $m \geq 0$, compact simple groups $\Gamma_1, \ldots, \Gamma_p$, and a finite covering homomorphism

$$\phi : \mathbb{K}^m \times \tilde{\Gamma}_1 \times \ldots \times \tilde{\Gamma}_p \to \Gamma,$$

Up to order, $\mathbb{K}^m$, $\Gamma_1, \ldots, \Gamma_p$ are uniquely determined by Γ.

If Γ is compact but not connected then the identity component Γ_0 of Γ is a normal subgroup of Γ and Γ/Γ_0 is finite.

Every closed subgroup of a Lie group has the structure of a Lie group. A compact Lie group has the (unique) structure of a real algebraic group. In particular, every compact Lie group can be realized as an algebraic subgroup of $\mathrm{SO}(n)$, for sufficiently large n.

Given $g_1, \ldots, g_N \in \Gamma$, let $\langle g_1, \ldots, g_N \rangle$ denote the closure (in Γ) of the subgroup generated by $g_1, \ldots, g_N$. If $\langle g_1, \ldots, g_N \rangle = \Gamma$, we say $g_1, \ldots, g_N$ are *topological generators* of Γ. If Γ is connected and abelian, then $\Gamma \approx \mathbb{K}^n$. It follows from Kronecker's theorem that $\mathbb{K}^n$ may be topologically generated by one group element [**14**]. Kronecker's theorem gives necessary and sufficient conditions for $\langle g \rangle = \mathbb{K}^n$ and it follows from these conditions that the topological generators of $\mathbb{K}^n$ form a full-measure subset with no interior points. If Γ is connected and non-abelian, we need two group elements to generate Γ and the set of generators forms a full-measure subset of Γ^2. If, in addition, Γ is semisimple, the set of pairs of group elements that topologically generate Γ forms a Zariski open subset of Γ^2 [**42, 27**].

Finally, every compact Lie group Γ admits a unique left- and right-translation invariant probability measure h – Haar measure. We often average over Haar measure to produce "Γ-invariant" objects.

2.1.2. Γ-manifolds. Let M be a connected smooth differential manifold. Denote the group of C^∞ diffeomorphisms of M by $\mathrm{Diff}(M)$. Suppose that Γ is a Lie group. A smooth Γ-action on M consists of a continuous momomorphism $\rho : \Gamma \to \mathrm{Diff}(M)$ such that the evaluation map $M \times \Gamma \to M; (x, g) \to \rho(g)(x)$ is smooth[1]. If we are given a smooth Γ-action on M, we say that M has the structure of a Γ-manifold. In future we set $\rho(g)(x) = gx$, $g \in \Gamma, x \in M$.

Henceforth we assume that Γ is compact.[2] If $x \in M$, let $\Gamma x = \{gx \mid g \in \Gamma\}$ denote the Γ-*orbit* of x. Every Γ-orbit is a compact Γ-invariant submanifold of M [**10**]. The *isotropy subgroup* of Γ at x is $\Gamma_x = \{g \in \Gamma \mid gx = x\}$. The isotropy subgroup is a closed, therefore Lie, subgroup of Γ. If $y = gx \in \Gamma x$ then $\Gamma_y = g\Gamma_x g^{-1}$ – that is, Γ_y and Γ_x are conjugate subgroups of Γ. We denote the conjugacy class of Γ_x in Γ by (Γ_x). If all isotropy subgroups are conjugate, we say the action of Γ is *monotypic*. Denote the identity element of Γ by e. The action of Γ on M is *free* if $\Gamma_x = \{e\}$ for all $x \in M$. Obviously, a free action

---

[1]Only continuity of the evaluation map is needed as smoothness follows from a result of Bochner-Montgomery [**48**, Chapter V].

[2]If Γ is not compact, many of the results we describe continue to hold under the assumption that the action of Γ is proper. That is, the map $(g, x) \mapsto (x, gx)$ is a proper mapping [**55**].

is monotypic. We say that the action of Γ is *transitive* if $\Gamma x = M$ for some (any) $x \in M$.

Let M/Γ denote the *orbit space* of M. That is, the quotient space of M by the relation $x \sim y$ if and only if x, y lie on the same Γ-orbit. While M/Γ is Hausdorff, M/Γ generally cannot be given the structure of a smooth manifold. However, if the action of Γ is free, M/Γ has the natural structure of a smooth manifold, the orbit map $p : M \to M/\Gamma$ is smooth and $p : M \to M/\Gamma$ has the structure of a principal Γ-bundle.

EXAMPLES 2.1.2. (1) Every Lie group Γ has the structure of a Γ-manifold with Γ-action defined by left-translation:

$$\rho_L(g)(\gamma) = g\gamma.$$

Right translation also gives Γ the structure of a Γ-manifold

$$\rho_R(h)(\gamma) = \gamma h^{-1}.$$

We can combine these actions to define a $\Gamma \times \Gamma$-action on Γ by

$$(g, h)\gamma = g\gamma h^{-1}.$$

All three actions are transitive. However, only the actions ρ_L, ρ_R are free.

(2) Let H be a closed subgroup of Γ. Then the *homogeneous space* Γ/H has the natural structure of a smooth Γ-manifold with the Γ action on cosets $\gamma[H]$ defined by left[3] multiplication: $\gamma[H] \mapsto g\gamma[H]$. The action of Γ is obviously transitive and not free (unless $H = \{e\}$). By restriction, any closed subgroup K of Γ defines a smooth K-action on Γ/H. This action will, in general, neither be free nor transitive. For example, if we embed SO(2) as the subgroup of SO(3) that fixes (say) the z-axis in $\mathbb{R}^3$, then SO(3)/SO(2) is diffeomorphic to the two-sphere S^2. The associated action of SO(3) on S^2 is the standard action defined by rotations of $\mathbb{R}^3$. If we take $K = \mathrm{SO}(2)$, then the action of K on S^2 is given by rotation about the z-axis. In this case, $K_{(0,0,\pm 1)} = \mathrm{SO}(2)$ and all other points of have trivial isotropy. If instead we take K to be the group order 24 consisting of the orientation preserving symmetries of the unit cube (with faces parallel to the coordinate planes and center at the origin), then possible isotropy groups for K are $\mathbb{Z}_2, \mathbb{Z}_3, \mathbb{Z}_4$ and the trivial group. The use of homogeneous spaces is a standard way to construct Γ-manifolds. For other methods, based on plumbing and geometric techniques we refer to [**10**].

(3) Let N be any smooth manifold. The product $N \times \Gamma$ has the natural structure of a Γ-manifold with Γ-action defined by left translation: $g(n, \gamma) = (n, g\gamma)$, $g \in \Gamma$, $(n, \gamma) \in N \times \Gamma$. Obviously the action of Γ

[3]We might also have considered right multiplication.

on $N \times \Gamma$ is free. More generally, the total space E of a Γ-principal bundle $\pi : E \to N$ has the structure of a Γ-manifold. In this case, the projection $\pi : E \to N$ is locally a product $U \times \Gamma$ and if $y \in E$, then $\pi^{-1}(\pi(y)) = \Gamma y$. This simple construction of Γ-manifolds plays an important role later when we come to construct skew product dynamical systems.
(4) Let H be a closed subgroup of Γ and suppose that N is a smooth H manifold. We define a free action of H on the product $N \times \Gamma$ by

$$h(n, \rho) = (hn, \rho h^{-1}), \quad (h \in H, \rho \in \Gamma,\ n \in N)$$

The *twisted product* $N \times_H \Gamma$ is the orbit space $(N \times \Gamma)/H$. Let $[n, \rho] \in N \times_H \Gamma$ denote the H-orbit through $(n, \rho) \in N \times \Gamma$. The twisted product $N \times_H \Gamma$ has the natural structure of a Γ-manifold with Γ-action defined by

$$g[n, \rho] = [n, g\rho], \quad (g, \rho \in \Gamma, n \in N).$$

Moreover, the orbit map $N \times \Gamma \to N \times_H \Gamma$ is an H-principal bundle over $N \times_H \Gamma$.

The action of H on N is not free if and only if the action of Γ on $N \times_H \Gamma$ is not free. If, for example, we let $\mathbb{Z}_2$ act on $(-1, 1)$ as multiplication by ± 1 and regard $\mathbb{Z}_2$ as the subgroup of $\mathbb{K}$ generated by $e^{\imath\pi}$, then $(-1, 1) \times_{\mathbb{Z}_2} \mathbb{K}$ is diffeomorphic to the Möbius band. The center circle of the Möbius band is a singular $\mathbb{K}$-orbit. Generally, twisted products give local models for Γ-manifolds. Indeed, every Γ-manifold is a finite union (with identifications) of twisted products (for more details on structural theorems for Γ-manifolds, see [**39**, Chapter 4]).
(5) Let $\rho : \Gamma \to \mathrm{O}(n)$ be a continuous homomorphism. Then we have an associated action of Γ on $\mathbb{R}^n$ by isometries defined by

$$(g, v) \mapsto \rho(g)(v), \quad g \in \Gamma, v \in \mathbb{R}^n.$$

This action, which is smooth, is called a *representation* of Γ on $\mathbb{R}^n$ (or just a Γ-representation of *degree* n). Any smooth action of Γ on $\mathbb{R}^n$ by linear maps can be regarded as a Γ-representation (average an inner product for $\mathbb{R}^n$ over Γ).
(6) Let M be a compact Γ-manifold. If we average a smooth riemannian metric for M over Γ using Haar measure, we may give M the structure of a smooth riemannian Γ-manifold. The Γ-action on M lifts to a Γ-action on the tangent bundle of M such that $g : T_xM \to T_{gx}M$ is a linear isometry for all $x \in M, g \in \Gamma$. Let $\alpha = \Gamma x$ be a Γ-orbit, $x \in M$. Denote the normal bundle of α by $q : N \to \alpha$. Then $q : N \to \alpha$ is Γ-vector bundle over α. In particular, each fiber N_x of N has the structure of a Γ_x-representation. If we fix $x \in \alpha$, there exists a Γ-vector bundle isomorphism $\chi : N \to N_x \times_{K_x} \Gamma$ defined by $\chi^{-1}([v, g]) = gv$,

$g \in \Gamma, v \in N_x$. As we indicate shortly, we can often reduce the study of dynamics in the neighborhood of an invariant Γ-orbit to dynamics on a twisted product. In turn, this dynamics can often be lifted to a skew product. These constructions play a very important role in equivariant dynamics. ♡

2.2. Equivariant dynamical systems

Suppose that Γ is a compact Lie group and M, N are smooth connected Γ-manifolds. A map $f : M \to N$ is Γ-*equivariant* or *equivariant* if

$$f(gx) = gf(x), \quad (x \in M, g \in \Gamma).$$

If the Γ-action on N is trivial, we say that a Γ-equivariant map $f : M \to N$ is G-*invariant* or just *invariant.*

EXAMPLES 2.2.1. (1) Suppose that $(\mathbb{R}^n, \Gamma)$ is a Γ-representation and $F : \mathbb{R}^n \to \mathbb{R}^n$ is Γ-equivariant. Then $fF : \mathbb{R}^n \to \mathbb{R}^n$ will be Γ-equivariant for all invariant $f : \mathbb{R}^n \to \mathbb{R}$.
(2) Let M be a Γ-manifold and $x \in M$. Then Γx is a Γ-invariant closed submanifold of M. The map $gx \mapsto g[\Gamma_x]$ defines a Γ-equivariant diffeomorphism between Γx and the homogeneous space Γ/Γ_x.
(3) If $f : M \to N$ is Γ-equivariant, then $\Gamma_{f(x)} \supset \Gamma_x$, all $x \in M$. If f is 1:1, then $\Gamma_{f(x)} = \Gamma_x$. If M is a group orbit and $f : M \to M$ is Γ-equivariant, then f is 1:1 and $f(y) \in N(\Gamma_y)(y)$, all $y \in \Gamma x$. ♡

2.2.1. Discrete dynamical systems.

Suppose that Γ is a compact Lie group and M is a compact Γ-manifold. Let $\text{Diff}^r_\Gamma(M)$ denote the group of C^r Γ-equivariant diffeomorphisms of M, $\infty > r > 0$, and $\text{Diff}_\Gamma(M)$ denote the group of C^∞ Γ-equivariant diffeomorphisms of M.

Let $f \in \text{Diff}^r_\Gamma(M)$ and suppose that $x \in M$ is a fixed point of f. It follows by equivariance that gx is also a fixed point of f, all $g \in \Gamma$, and so Γx is fixed for f. We briefly consider the issue of the stability of f-invariant orbit Gx under perturbations of f in $\text{Diff}^r_\Gamma(M)$, $r \geq 1$ (for more details see [**20, 25, 26**]). If Γ is finite, then Γx will be a finite set of fixed points and the natural condition to require of x is that x is a hyperbolic fixed point of f. It is easy to verify that x is a hyperbolic fixed point if and only if gx is a hyperbolic fixed point, $g \in \Gamma$. When Γ is not finite, more precisely if Γ_x is of dimension strictly less than Γ, then points on Γx can never be hyperbolic. Further, it is no longer reasonable to expect that *fixed* points will generally persist under perturbation by f. Specifically, let $N(\Gamma_x)$ denote the normalizer of Γ_x in Γ and r_x denote the rank[4] of the group $N(\Gamma_x)/\Gamma_x$. If $r_x > 0$, we can

[4]The dimension of a maximal torus in the identity component of $N(\Gamma_x)/\Gamma_x$.

make arbitrarily small smooth perturbations f' of f along Γx such that f' has no fixed points on Γx even though $f'(\Gamma x) = \Gamma x$. For this reason, when Γ is infinite, it is more natural to study f-invariant Γ-orbits rather than orbits of fixed points. The natural stability condition to require of f-invariant Γ-orbits is *normal hyperbolicity*. It is not hard to show that invariant normally hyperbolic Γ-orbits persist under perturbation of f. Moreover, if Γx is an invariant Γ-orbit of f and $f(x) = gx$, then Γx is normally hyperbolic if and only if the spectrum of $D(g^{-1}f)(x)$ contains exactly $\dim(\Gamma x)$ eigenvalues of unit modulus [**20, 25**]. We may similarly develop the stability theory of periodic invariant Γ-orbits. Of course, if we let $\tilde{f} : M/\Gamma \to M/\Gamma$ denote the homeomorphism induced by f on the orbit space, then a Γ-orbit α will be an invariant periodic orbit for f if and only if $\alpha \in M/\Gamma$ is a periodic point of $\tilde{f}$.

2.2.2. Skew and principal extensions. Let $f : N \to N$ be a C^r diffeomorphism (no equivariance). A C^r*-cocycle* on N is a C^r-map $\beta : N \to \Gamma$. Let $C^r(N,\Gamma)$ denote the group of C^r-cocycles on N. Associated to every $\beta \in C^r(N,\Gamma)$, we may define $f_\beta \in \mathrm{Diff}^r_\Gamma(N \times \Gamma)$ by

$$f_\beta(n,g) = (f(n), g\beta(n)), \quad (n \in N, g \in \Gamma).$$

We refer to f_β as a Γ*-extension of* f. In this case, every fixed or periodic point of f gives rise to an invariant or periodic invariant F-orbit of f_β. Hyperbolicity of fixed or periodic points of f is equivariant to normal hyperbolicity of the corresponding invariant Γ-orbits of f_β.

If $\pi : E \to N$ is a Γ-principal bundle over N, then any $F \in \mathrm{Diff}^r_\Gamma(E)$ which covers f ($\pi F = f\pi$) is called a *principal* Γ*-extension* of f. We define the C^r cocycles to be the group $C^r_\Gamma(E,\Gamma)$ of C^r maps $\beta : E \to \Gamma$ such that $\beta(gx) = g\beta(x)g^{-1}$, $g \in \Gamma, x \in E$. Given $F \in \mathrm{Diff}^r_\Gamma(E)$, $\beta \in C^r_\Gamma(E,\Gamma)$, we define $F_\beta \in \mathrm{Diff}^r_\Gamma(E)$ by $F_\beta(x) = \beta(x)F(x)$, $x \in E$.

2.2.3. Continuous dynamical systems. Since continuous equivariant dynamical systems do not play a major role in this work, we give minimal details. If M is a compact Γ-manifold, then the Γ-action on M lifts to a Γ action on the tangent bundle TM of M. Let $C^r_\Gamma(TM)$ denote the space of C^r Γ-equivariant vector fields on M, $r \geq 0$. If $X \in C^r_\Gamma(TM)$, $r \geq 1$, we let Φ^X_t denote the corresponding flow on M. It follows from the Γ-equivariance of X and the uniqueness theorem for differential equations that $\Phi^X_t \in \mathrm{Diff}^r_\Gamma(M)$, all $t \in \mathbb{R}$.

2.3. Local theory

In this section, we restrict mainly to discrete dynamical systems. Suppose that $\alpha = \Gamma x$ is an invariant Γ-orbit for $f \in \mathrm{Diff}^r_\Gamma(M)$. Choose a G-invariant riemannian metric on M and identify the normal bundle

N of α with the orthogonal complement of $T\alpha$ in TM. It follows from the equivariant version of the tubular neighborhood theorem (see [**10**, Chapter VI]), that there exists a Γ-equivariant diffeomorphism q of N onto an open Γ invariant neighborhood U of α in M. The normal bundle N is Γ-vector bundle isomorphic to the twisted product $N_x \times_{\Gamma_x} \alpha$ ($N_x \times_{\Gamma_x} \alpha$ has the structure of a Γ-vector bundle over α). It follows that for the local study of the dynamics on f near α, we can restrict attention to equivariant diffeomorphisms F of $N_x \times_{\Gamma_x} \alpha$ which cover the map $f : \alpha \to \alpha$. A further simplification is obtained by noting that since F is equivariant, it is uniquely determined by its values on a fiber of $N_x \times_{\Gamma_x} \alpha$, say N_x. If we set $S_y = q(N_y)$, $y \in \alpha$, then S_y is a differentiable *slice* for the action of Γ at y[5]. Again, the values of f on U are uniquely determined by $f|S_y$, any $y \in \alpha$. Typically, we do perturbation theory for f near α by restricting f to a slice S, perturbing $f|S$ to f' and then extending f' Γ-equivariantly to U. In this way, for example, we can prove that invariant Γ-orbits (or periodic Γ-orbits) are generically normally hyperbolic. We remark that for flows, we can always lift dynamics on the twisted product $N_x \times_{\Gamma_x} \alpha$ to the skew extension $N_x \times \alpha$. Generally, this is not possible for discrete dynamical systems unless the diffeomorphism f is equivariantly isotopic to the identity map. We refer to [**20, 25**] for more details as well as a statement of a Kupka-Smale density theorem. We remark that the theory for vector fields and flows is somewhat easier than that for discrete dynamical systems.

2.4. Invariant subspaces and transversality

If the action of Γ on M is free (M is a skew or principal Γ-extension over M/Γ), then M/Γ is a smooth manifold and every discrete or continuous smooth dynamical system on M drops down to a smooth dynamical system on the orbit space M/Γ. Conversely, smooth flows on M/Γ always lift to smooth Γ-equivariant flows on M. For discrete systems, there may be topological obstructions to lifting from M/Γ to M. In any event, the only new features seen in equivariant dynamics on M come from the fact that there may be 'drift' along Γ-orbits in M. Thus the geometric theory of equivariant dynamics on M can effectively be reduced to the study of the way the 'base' dynamics on M/Γ interacts with the effect of drift along Γ-orbits. For this reason Lie group extensions of prescribed base dynamics are frequently used as mathematical

[5]That is, S_y is Γ_y-invariant, $S_x \cap S_y \neq \emptyset$ if and only if $x = y$, $\gamma S_y = S_{\gamma y}$, $\gamma \in \Gamma$, and ΓS_x is an open neighborhood of α. Since S_y is Γ_y-equivariantly diffeomorphic to N_y, we can assume the action of Γ_y on S_y is linear [**10**, Chapter VI].

models of physical systems with Lie group symmetry. If the action of Γ on M is monotypic, then let H be an isotropy group of the action and let M^H denote the fixed point set of the action of H on M. Then $K = N(H)/H$ acts freely on the compact manifold M^H and we reduce to the previous case.

Matters are different if the action of Γ is not monotypic. Let $\mathbf{I} = \mathbf{I}(M)$ denote the set of isotropy groups for the action of Γ on M. If $H \in \mathbf{I}$, let $M_H = \{x \in M \mid \Gamma_x = H\}$. Then M_H is a submanifold of M which is invariant by all Γ-equivariant diffeomorphisms and flows on M. Consequently, $\{M_H \mid H \in \mathbf{I}\}$ defines a partition of M into submanifolds invariant by all Γ-equivariant diffeomorphisms and flows. Typically, the *principal* (minimal) isotropy group of a Γ-action is (e). The corresponding Γ-invariant set, which we denote M_N, is then an open and dense submanifold of M. If Γ is connected or Γ contains no reflections, then M_N is connected [**10**]. Γ-orbits not in M_N are called *singular*.

EXAMPLE 2.4.1. Let $\mathbf{C}$ be the group of orientation preserving symmetries of the unit cube in $\mathbb{R}^3$. Points lying on the coordinate axes have isotropy group in $(\mathbb{Z}_4)$, on axes through opposite vertexes have isotropy group in $(\mathbb{Z}_3)$, on axes through midpoints of opposite edges have isotropy group in $(\mathbb{Z}_2)$, the origin has isotropy group $\mathbf{C}$, and all other points have trivial isotropy. Every $\mathbf{C}$-equivariant flow on $\mathbb{R}^3$ leaves all axes as well as the origin fixed. A consequence is that the origin has to be an equilibrium point and if the origin is hyperbolic then it must either be a sink or source. ♡

In general, the presence of subspaces invariant by all dynamical systems implies the possibility of stably non-transverse intersections of invariant manifolds. All of this is handled by equivariant transversality theory [**4, 5, 19, 26**]. Although we shall not make use of this theory here, it is worth pointing out that one important feature of equivariant dynamics is the possibility of *structurally stable* heteroclinic cycles of non-transverse intersections of invariant manifolds of fixed or period points [**34**]. In simple cases, heteroclinic cycles can be attractors. Attractors of this form stably exhibit non-ergodic averages. This feature of equivariant dynamics should be contrasted with the expected generic behavior in asymmetric systems (cf the *Palis conjecture* [**56**]).

2.5. Basic sets for equivariant diffeomorphisms

The main aim of this work is to study basic sets of Γ-equivariant diffeomorphisms that extend the class of skew and principal extensions. In particular, we want to allow the possibility that Γ does not act freely (or monotypically) on the basic set. In [**28**], we constructed transitive hyperbolic attractors for diffeomorphisms and flows equivariant with respect to a finite group. In that work, attractors were invariant by a subgroup H of Γ which always acted freely on the attractor. In other work [**22**], we have constructed geometric realizations of Γ-invariant subshifts of finite type. For these constructions we allowed Γ to be a general compact group and we also permitted at least some singular Γ-orbits. Even when Γ is finite, the presence of singular orbits is often associated to the presence of non-uniformly hyperbolic structure. This is seen in the $\mathbb{Z}_2$-symmetric Lorenz attractor and in attractors related to solenoids [**31**]. In the present work we focus on uniform hyperbolic behavior in the presence of singular Γ-orbits.

CHAPTER 3

Technical preliminaries

3.1. Geometry of group actions and maps

Henceforth, Γ will always denote a compact Lie group with identity e. If H is a (closed) subgroup of Γ, let $N(H)$ and $C(H)$ respectively denote the normalizer and centralizer of H in Γ. Let d_Γ be a metric on Γ which is invariant under left and right translations by elements of Γ (see Examples 2.1.2(1) and average any smooth riemannian metric on Γ using Haar measure for $\Gamma \times \Gamma$).

3.1.1. Isotropy types and strata. We recall some facts about orbit strata for Γ-actions [**71**]. Let M be a smooth Γ-manifold. Given $x \in M$, let (T_xM, Γ_x) denote the representation of $\Gamma_x \in \mathbf{I}$ induced on the tangent space T_xM of M at x. Since $T_x\Gamma x$ is Γ_x-invariant, we have an associated representation (N_x, Γ_x) of Γ_x on the normal space $N_x = T_xM/T_x\Gamma x$ at x to Γx. We define an equivalence relation $\sim$ on the set of all representations (N_x, Γ_x), $x \in M$, by $(N_x, \Gamma_x) \sim (N_y, \Gamma_y)$ if and only if Γ_x and Γ_y are conjugate subgroups of Γ and the representations (N_x, Γ_x) , (N_y, Γ_y) are isomorphic. (If M is a Γ-representation, then $(N_x, \Gamma_x) \sim (N_y, \Gamma_y)$ if and only if Γ_x, Γ_y are conjugate subgroups of Γ.) We call an equivalence class of $\sim$ an *isotropy type* for the action of Γ on M and let $\mathcal{O} = \mathcal{O}(M, \Gamma)$ denote the set of isotropy types. We occasionally abuse notation and write "$H \in \tau$" to mean that for some $x \in M$, $\Gamma_x = H$ and $(N_x, \Gamma_x) \in \tau$ (there is no ambiguity if M is a Γ-representation). If M is compact or a representation, then $\mathcal{O}$ is finite. Given $\tau \in \mathcal{O}$, let M_τ denote the set of points with isotropy type τ. We refer to M_τ as an *orbit stratum* and note that M_τ is a smooth Γ-invariant submanifold of M.

If $H \in \mathbf{I}$, then M^H (the fixed point set of H) and M_H (the subset of M^H with isotropy group H) have the structure of smooth $N(H)$-manifolds.

EXAMPLE 3.1.1. If (V, Γ) is a Γ-representation and $H \in \mathbf{I}$, then V_H is an open and dense subset of V^H. Indeed, let $v \in V_H$ and choose a slice S_v for the action of Γ at v. Then for all $w \in S_v$, $\Gamma_w \subset H$ and so $\Gamma_w = H$ for all $w \in S_v \cap V^H$. This proves that V_H is an open subset of

V^H. Density is now easy if Γ is a finite group ($V_H = V^H \setminus \cup_{J\in \mathbf{I}, J\supsetneq H} V^J$). If Γ is infinite, let $u \neq v \in V_H$ and $\ell \subset V^H$ be the line through u, v. If $w \in \ell \cap \partial V_H$, then it follows by taking a slice at w that points of ℓ near, but not equal to w, have isotropy H. It follows that $\ell \cap V_H$ is open and dense in ℓ and hence that V_H is open and dense in V^H.

If M is a Γ-manifold and $H \in \mathbf{I}$, then M_H is always a non-empty open subset of M^H. However, M_H will generally *not* be a dense subset of M^H. ♡

Let $\tau \in \mathcal{O}$. Since $M_H \subset M^H$ and $\Gamma M^H = \cup_{H\in\tau} M^H$ is closed, it follows that

$$\overline{M_\tau} \subset \cup_{H\in\tau} M^H. \tag{3.1}$$

REMARK 3.1.2. We have equality in (3.1) for Γ-representations (V, Γ) (since by Example 3.1.1, $V_\tau = \Gamma V_H$, $H \in \tau$). ♢

EXAMPLES 3.1.3. (1) Let $\mathbf{C}$ denote the group of orientation preserving symmetries of the cube (see Example 2.4.1), Let $\tau = (\mathbb{Z}_4)$. In this case $\mathbb{R}^3_\tau$ consists of the of the three coordinate axes, less the origin, and so $\mathbb{R}^3_\tau = \cup_{H\in\tau} (\mathbb{R}^3)^H \setminus \{0\}$, where the union is over the three conjugate copies of $\mathbb{Z}_4$ in $\mathbf{C}$.
(2) Let $\mathbb{K}$ act on the unit sphere $S^3 \subset \mathbb{C}^2$ according to $\theta(z_1, z_2) = (e^{\imath p\theta} z_1, e^{\imath q\theta} z_2)$, where p, q are strictly positive integers. If $p = q = 1$, $\mathbb{K}$ acts freely on S^3 and we recover the Hopf fibration $S^3 \to S^3/\mathbb{K} = S^2$. If $p = 1$, $q > 1$, the action on S^3 has one singular $\mathbb{K}$-orbit with isotropy type $(\mathbb{Z}_p)$. If $(p, q) = 1$, $p, q > 1$, then the action on S^3 has two singular $\mathbb{K}$-orbits with isotropy types $(\mathbb{Z}_p), (\mathbb{Z}_q)$. Neither of these non-free actions on S^3 can be represented as twisted products of $\mathbb{K}$ with a compact surface. ♡

We define a partial order on $\mathcal{O}$ by $\tau < \eta$ if $\partial M_\tau \cap M_\eta \neq \emptyset$. If $\tau < \eta$, it follows, using slices, that there exist $H \in \tau$, $J \in \eta$ such that $H \subsetneq J$.

REMARK 3.1.4. If (V, Γ) is a Γ-representation, then $\tau < \eta$ if and only if there exist $H \in \tau$, $J \in \eta$ such that $H \subsetneq J$. This is not the case for general Γ-manifolds. ♢

If M is connected, there exists a unique minimal isotropy type Π for the order on $\mathcal{O}$ – the *principal* isotropy type. The principal orbit stratum M_Π is open and dense in M and connected if there are no orbit strata of codimension 1 in M. Let $M_S = M \setminus M_\Pi$ and refer to orbits in M_S as *singular* orbits.

REMARK 3.1.5. Suppose that M is connected. If Γ is finite, it follows from the differentiable slice theorem that Π consists of a unique pair (P, m), where P is a normal subgroup of Γ and m is a trivial P-representation. Since P must fix all points in M_Π, it follows that P acts trivially on M. Hence, it is no loss of generality to replace Γ by Γ/P and assume that Γ acts *freely* on M_Π. If Γ is infinite, then Π may contain infinitely many distinct isotropy groups. Suppose that $(P, m) \in \Pi$ and consider M^P. Since P is a normal subgroup of $N(P)$, $N(P)/P$ acts freely on $M_P \subset M^P$. In general, it is not true that smooth (at least C^1) $N(P)/P$-equivariant vector fields (resp diffeomorphisms) on M^P extend to smooth Γ-equivariant vector fields (resp diffeomorphisms) on M (for an example, see [**71**, Example 11.10]). However, if we assume that all Γ-orbits have the same dimension, then it is easy to show that every smooth $N(P)/P$-equivariant vector field (resp diffeomorphism) on M^P extends uniquely to a smooth Γ-equivariant vector field (resp diffeomorphism) on M. Typically we shall be studying compact Γ-invariant subsets Λ of M such that the Γ-orbits of points in Λ all have the same dimension, say $d \leq \dim(\Gamma)$. If $d \neq \dim(\Gamma)$, we can replace M by $M(d)$ – the subset of M consisting of all points x such that $\dim(\Gamma x) = d$. The previous considerations then apply to the principal stratum of $M(d)$. As a consequence of these remarks, we will usually assume in the sequel that Γ acts *freely* on the principal orbit stratum and that all isotropy groups for the action are *finite*. ◊

Following [**26**, §9.3], we define a filtration of M associated to the order on $\mathcal{O}$.

DEFINITION 3.1.6. An isotropy type $\tau \in \mathcal{O}$ is *k-submaximal* if k is the largest integer such that there exist isotropy types $\eta_1, \ldots, \eta_k \in \mathcal{O}$ satisfying (a) η_1 is maximal; (b) $\eta_k = \tau$; (c) $\eta_1 > \ldots > \eta_k$.

Let $\mathcal{M}_k = \mathcal{M}_k(M)$ denote the set of k-submaximal isotropy types and N be the largest integer such that $\mathcal{M}_N \neq \emptyset$. The set $\mathcal{M}_1$ consists of the maximal isotropy types and, provided M is connected, $\mathcal{M}_N$ consists of the principal isotropy type Π. For $k \geq 1$, define $M_k = \cup_{\tau \in \mathcal{M}_k} M_\tau$, $M^k = \cup_{i=1}^k M_i$. Set $M_0 = M^0 = \emptyset$. The next result is a simple consequence of our definitions and slice theory.

LEMMA 3.1.7. (1) $M^1 \subsetneq \ldots \subsetneq M^N = M$.
(2) *M^j is a closed Γ-invariant subset of M, $1 \leq j \leq N$.*
(3) *M_j is a smooth Γ-invariant submanifold of M, $1 \leq j \leq N$.*
(4) $\partial M_j \subset M^{j-1}$, $1 \leq j \leq N$.

Our notation extends naturally to the case when Λ is a Γ-invariant subset of M. In this case, however, Λ may not have a unique minimal isotropy type, even if Λ contains points of M-principal isotropy. If Λ

is closed, then parts (1,2,4) of Lemma 3.1.7 hold if we replace M_j by $\Lambda_j = M_j \cap \Lambda$ and M^j by $\Lambda^j = M^j \cap \Lambda$.

3.1.2. Orbit structure of twisted products. Let H be a closed subgroup of Γ and suppose that Z is an H-space. Form the twisted product $\Lambda = Z \times_H \Gamma$ and note that Λ has the structure of a Γ-space. Of course, if Z is an H-manifold, Λ is a Γ-manifold (see Examples 2.1.2(3)).

If $x = [z, \gamma] \in Z \times_H \Gamma$, then $\Gamma_x = \gamma H_z \gamma^{-1}$, where H_z is the isotropy group at z for the action of H on Z. It follows that we can identify $\mathcal{O}(Z, H)$ with $\mathcal{O}(Z \times_H \Gamma, \Gamma)$. With this identification, we have $(Z \times_H \Gamma)_\tau \approx Z \times_H \Gamma_\tau$ for all $\tau \in \mathcal{O}$.

Let $S = \{[z, e] \mid z \in Z\}$. Then S is an H-invariant subset of $Z \times_H \Gamma$. Every Γ-orbit of a point in $Z \times_H \Gamma$ meets S in a unique H-orbit. If every H orbit is finite, each intersection with S will consist of a finite set of points and every Γ-orbit will have the same dimension.

We use the following elementary result repeatedly in the sequel.

LEMMA 3.1.8. *Let H be a closed subgroup of Γ and suppose that Z is an H-space. The natural H-equivariant inclusion $Z \subset \Gamma \times_H Z$ induces a natural homeomorphism $Z/H \approx (\Gamma \times_H Z)/\Gamma$.*

3.1.3. Equivariant vector fields. For simplicity of notation, we usually assume that diffeomorphisms are smooth – that is, C^∞. However, all of the results in Part I hold under the assumption that diffeomorphisms are C^s, $\infty \geq s \geq 1$. In Part II, we frequently need to assume that diffeomorphisms are of class C^s, $s \geq 2$.

Let $C^0_\Gamma(TM)$ and $C^\infty_\Gamma(TM)$ respectively denote the spaces of continuous and smooth Γ-equivariant vector fields on M. If Λ is a Γ-invariant compact subset of M, let $C^0_\Gamma(T_\Lambda M)$ denote the space of continuous Γ-equivariant sections of $T_\Lambda M = TM|\Lambda$. Let $C^\infty_\Gamma(T_\Lambda M)$ denote the subspace of $C^0_\Gamma(T_\Lambda M)$ consisting of vector fields which extend to smooth vector fields on M.

For each $x \in M$, let $\mathbf{T}_x = T_x\Gamma x$ and set $\mathbf{T} = \cup_{x \in M} \mathbf{T}_x$. Obviously, $\mathbf{T}$ is a Γ-invariant subspace of TM. If $f \in \mathrm{Diff}_\Gamma(M)$, $Tf : \mathbf{T} \to \mathbf{T}$. Let $d \in \mathbb{N}$ and define $M(d) = \{x \in M \mid \dim(\Gamma x) = d\}$. Using slices, one may easily verify that $M(d)$ is a smooth Γ-invariant submanifold of M (use [**22**, Lemma A, §3]). Clearly, $\mathbf{T}|M(d)$ is a smooth Γ-subbundle of $TM|M(d)$. More generally, if $\Lambda \subset M(d)$ is Γ-invariant, then $\mathbf{T}_\Lambda = \mathbf{T}|\Lambda$ has the structure of a (continuous) Γ-subbundle of $T_\Lambda M$. Set $\mathcal{T}(M) = \{X \in C^\infty_\Gamma(TM) \mid X(x) \in \mathbf{T}_x,\ x \in M\}$. (Vector fields in $\mathcal{T}(M)$ are everywhere tangent to Γ-orbits.) If $\Lambda \subset M$ is a closed Γ-invariant set, let $\mathcal{T}^0_\Lambda(M)$ denote the space of continuous Γ-equivariant sections of $\mathcal{T}|\Lambda$ and $\mathcal{T}_\Lambda(M)$ denote the subspace of $\mathcal{T}^0(\Lambda)$ consisting of vector fields which are restrictions of elements of $\mathcal{T}(M)$.

Suppose that $f \in \mathrm{Diff}_\Gamma(M)$. Let $\tilde{f} : M/\Gamma \to M/\Gamma$ denote the map induced by f on the orbit space M/Γ. Define

$$\mathcal{D}_f(M) = \{g \in \mathrm{Diff}_\Gamma(M) \mid \tilde{g} = \tilde{f}\}$$

Obviously, if $g, h \in \mathcal{D}_f(M)$, then $gh^{-1} \in \mathcal{D}_I(M)$, where $I = I_M$ denotes the identity map of M. Note that $f \in \mathcal{D}_I(M)$ if and only if $f(x) \in \Gamma x$, all $x \in M$.

3.1.4. Equivariant maps of twisted products. Let H be a closed subgroup of Γ and Z be an H-space. Let H act on Γ via conjugation ($h(\gamma) = h\gamma h^{-1}$, $h \in H$, $\gamma \in \Gamma$). If $f : Z \to Z$ and $\phi : Z \to \Gamma$ are continuous and H-equivariant, then we may define the continuous Γ-equivariant map $f_\phi : Z \times_H \Gamma \to Z \times_H \Gamma$ by

$$f_\phi([z, \gamma]) = [f(z), \gamma\phi(z)], \quad ((z, \gamma) \in Z \times \Gamma).$$

Of course, f_ϕ is just induced from the skew extension of f to $\Gamma \times Z$ with cocycle ϕ. At the orbit space level, we have $\widetilde{f_\phi} = \tilde{f}$, where $\widetilde{f_\phi}$ is the map induced by f_ϕ on $(Z \times_H \Gamma)/\Gamma$ and $\tilde{f}$ the map induced by f on Z/H.

Suppose now that P is a smooth H-manifold and that all H-orbits are finite. Set $P \times_H \Gamma = M$. Let $f \in \mathrm{Diff}_H(P)$. Define $\mathbf{e} \in C^\infty_H(P, \Gamma)$, by $\mathbf{e}(p) = e$, $p \in P$. If we define $I_\phi \in \mathrm{Diff}_\Gamma(M)$ by

$$I_\phi([x, \gamma]) \quad = \quad [x, \gamma\phi(x)], \quad ((x, \gamma) \in P \times \Gamma),$$

then we have the decomposition $f_\phi = I_\phi \circ f_{\mathbf{e}}$, for all cocycles $\phi \in C^\infty_H(P, \Gamma)$. Obviously, $\widetilde{f_\phi} = \widetilde{f_{\mathbf{e}}}$ and $\widetilde{I_\phi}$ is the identity map of P/H.

LEMMA 3.1.9 ([**21**],[**22**, Lemma C]). *Let $\phi \in C^\infty_H(P, \Gamma)$. There is an open neighborhood $\mathcal{U}$ of f_ϕ in $\mathcal{D}_f(M)$ and a continuous (C^∞-topology) map $\rho : \mathcal{U} \to C^\infty_H(P, \Gamma)$ such that for all $h \in \mathcal{U}$, $h = I_{\rho(u)} \circ f$.*

REMARKS 3.1.10. (1) We may take $\mathcal{U}$ to be the set consisting of diffeomorphisms isotopic to f_ϕ within $\mathcal{D}_f(M)$ [**22**, Lemma D]. (3) Lemma 3.1.9 fails if Γ-orbits are not of constant dimension. ♢

Using slices, we have the following straightforward application of Lemma 3.1.9.

PROPOSITION 3.1.11. *Let M be a compact Γ manifold and suppose that all Γ-orbits are of the same dimension. Given $f \in \mathrm{Diff}_\Gamma(M)$, there exists an open neighborhood $\mathcal{U}$ of $f \in \mathcal{D}_f(M)$ and a continuous (C^∞-topologies) map $\phi : \mathcal{U} \to C^\infty_H(P, \Gamma)$ such that for all $u \in \mathcal{U}$, $u = I_{\phi(u)} \circ f$.*

3.2. Equivariant subshifts of finite type: Γ finite

Throughout this section we assume Γ is a finite group. Some of what we discuss here is presented in greater detail in [**22**, §2].

3.2.1. Subshifts of finite type. For $n \geq 1$, let $\mathbf{n} = \{1, \ldots, n\}$ and S_n denote the associated symmetric group. Let $\mathbf{M}(n) = \mathbf{M}$ denote the set of $n \times n$ 01 matrices. If $A \in \mathbf{M}$, we let $\sigma : \Sigma_A \to \Sigma_A$ denote the associated subshift of finite type. Topologize Σ_A as a subspace of $\mathbf{n}^{\mathbb{Z}}$ (Tychonov product topology).

Suppose that $\psi : \Gamma \to S_n$ is a permutation representation of Γ. Associated to ψ we have actions of Γ on $\mathbf{M}$ and $\mathbf{n}^{\mathbb{Z}}$ defined by

$$\begin{aligned} \gamma(A)(i,j) &= A(\psi(\gamma)(i), \psi(\gamma)(j)), \quad (\gamma \in \Gamma,\ A \in \mathbf{M},\ i,j \in \mathbf{n}), \\ \gamma(x)_i &= \psi(\gamma)(x_i), \quad x \in \mathbf{n}^{\mathbb{Z}},\ -\infty < i < \infty. \end{aligned}$$

Let $\mathbf{M}^{\Gamma}$ denote the fixed point set of this action and observe that if $A \in \mathbf{M}^{\Gamma}$ then Σ_A is a Γ-invariant subset of $\mathbf{n}^{\mathbb{Z}}$. The next lemma is a trivial consequence of our definitions.

LEMMA 3.2.1. *Let $A \in \mathbf{M}$. Then $\sigma : \Sigma_A \to \Sigma_A$ is a Γ-equivariant homeomorphism if and only if $A \in \mathbf{M}^{\Gamma}$.*

REMARK 3.2.2. It is shown in [**1**] that every subshift of finite type which admits a finite group action Γ commuting with the shift map is conjugate to a subshift $\sigma : \Sigma_A \to \Sigma_A$ of finite type with $A \in M^{\Gamma}$. $\diamondsuit$

EXAMPLE 3.2.3. Let Γ be a subgroup of S_n. If the action of Γ on $\mathbf{n}$ is not free (that is, there exists $j \in \mathbf{n}$ such that $\Gamma_j \neq \{e\}$), then $\#\mathcal{O}(\mathbf{n}^{\mathbb{Z}}, \Gamma) \geq 2$. Otherwise, Γ acts freely on $\mathbf{n}^{\mathbb{Z}}$. First observe that a point in $\mathbf{n}^{\mathbb{Z}}$ containing all symbols always has trivial isotropy. If $\Gamma_j \neq \{e\}$, then $\bar{j} \cdot \bar{j} \in \mathbf{n}^{\mathbb{Z}}$ has isotropy $\Gamma_j \neq \{e\}$ and so $\#\mathcal{O}(\mathbf{n}^{\mathbb{Z}}, \Gamma) \geq 2$. On the other hand, every point $(x_i) \in \mathbf{n}^{\mathbb{Z}}$ has trivial isotropy, if and only if $\Gamma_j = \{e\}$, all $j \in \mathbf{n}$. In this case, Γ acts freely on $\mathbf{n}^{\mathbb{Z}}$ and so $\#\mathcal{O}(\mathbf{n}^{\mathbb{Z}}, \Gamma) = 1$. $\heartsuit$

EXAMPLE 3.2.4. Set $X = \mathbf{n}^{\mathbb{Z}}$ and suppose that $\Gamma \subset S_n$ acts freely and transitively on $\mathbf{n}$. Clearly, $|\Gamma| = n$ and Γ acts freely on $\mathbf{n}^{\mathbb{Z}}$. Let $\tilde{\sigma} : X/\Gamma \to X/\Gamma$ denote the map induced on the orbit space by σ. Then there is a homeomorphism $h : X/\Gamma \to X$ conjugating σ and $\tilde{\sigma}$. A similar result holds for subshifts $\sigma : \Sigma_A \to \Sigma_A$ of finite type provided that $A \in M^{\Gamma}$. The proof amounts to a re-coding of X. Specifically, set $n^2 = p$ and let $\mathbf{p}$ denote the set of all pairs (i,j), $i,j \in \mathbf{n}$. We may embed X as a subshift of finite type in $\mathbf{p}^{\mathbb{Z}}$ using the natural embedding $\iota : \mathbf{n}^{\mathbb{Z}} \to \mathbf{p}^{\mathbb{Z}}$ defined by $\iota((x_i)) = ((x_i x_{i+1}))$. Let A denote the associated 01 matrix. Obviously, $\iota\sigma = \sigma'\iota$, where σ' denotes the shift on $\mathbf{p}^{\mathbb{Z}}$. The action of Γ on $\mathbf{n}$ extends to a free action on $\mathbf{p}$ (and so also on $\mathbf{p}^{\mathbb{Z}}$). Let $p : \mathbf{p} \to \tilde{\mathbf{n}}$ denote the orbit map. Since Γ acts freely on $\mathbf{p}$, $|\tilde{\mathbf{n}}| = n$. Define $\rho : X \to \tilde{\mathbf{n}}^{\mathbb{Z}}$ by $\rho((x_i)) = (p(x_i x_{i+1}))$. Obviously ρ is continuous, surjective, Γ-invariant and commutes with the shift maps. Hence ρ induces a continuous surjection $h : X/\Gamma \to \tilde{\mathbf{n}}^{\mathbb{Z}}$ such that $h\tilde{\sigma} = \sigma h$. It remains to prove that h is injective. It follows from the definition of ι that for each $(i,j) \in \mathbf{p}$, $A((i,j),(i',j')) = 1$, if and only if $i' = j$. Suppose $z \in \tilde{\mathbf{n}}^{\mathbb{Z}}$ and choose $y = (y_i) \in \iota(X)$ such that $\rho(y) = z$. Acting by an element of Γ, we may assume that $y_0 = (1,j)$. But then y_1 is determined as the unique pair (i',j') such that $p(i',j') = z_1$ and $i' = j$. Proceeding inductively, we see that the y_i are uniquely determined once we have made a choice for y_0. It follows that for all $z \in \tilde{\mathbf{n}}^{\mathbb{Z}}$, $\rho^{-1}(z)$ consists of precisely one Γ-orbit and so h is injective. As pointed out to us by the referee, there is a much simpler proof if Γ is abelian: define the Γ-invariant map $h : \mathbf{n}^{\mathbb{Z}} \to \Gamma^{\mathbb{Z}}$ by $h((x_i)) = (\gamma_i)$, where γ_i is the unique element of Γ such that $\gamma_i = x_{i+1}$. ♡

DEFINITION 3.2.5 ([**22**, §2]). Suppose that Σ is a compact metric Γ-space and $\alpha : \Sigma \to \Sigma$ is a Γ-equivariant homeomorphism. We say that the pair (Σ, α) is a Γ-subshift of finite type if, for some $n \geq 1$, there exist a representation ψ of Γ in S_n, $A \in \mathbf{M}(n)^{\Gamma}$ and an equivariant homeomorphism $h : \Sigma_A \to \Sigma$ such that $\alpha = h \circ \sigma \circ h^{-1}$.

3.3. Hölder continuity and the Ruelle operator

We recall some standard definitions and properties about subshifts of finite type. For proofs and more details we refer to Bowen [**9**], Parry & Pollicott[**58**, Chapter 1], or Katok & Hasselblatt [**38**, Chapter 1]. Henceforth, we shall assume A is irreducible. Let d the period of A and recall that A is *aperiodic* if $d = 1$.

LEMMA 3.3.1. *Suppose $A \in \mathbf{M}^{\Gamma}$ is irreducible with period d. There is a partition $X_1 \cup \ldots \cup X_d$ of Σ_A into open and closed sets such that for $1 \leq \ell \leq d$*

(a) $\sigma(X_\ell) = X_{\ell'}$, $(\ell' \equiv \ell + 1 \mod d)$.
(b) $\sigma^d | X_\ell$ *corresponds to an aperiodic matrix.*
(c) Γ *acts as a permutation group on* $\{X_1, \ldots, X_d\}$.

REMARK 3.3.2. It is easy to find examples to show that Γ can act trivially or non-trivially on $\{X_1, \ldots, X_d\}$. ♢

EXAMPLE 3.3.3. Suppose that $T = \psi(\Gamma)$ is a transitive subgroup of S_n. Let $A \in \mathbf{M}$ and suppose that at least one non-diagonal entry of

A is not zero. Then A is irreducible. If, in addition, A has a non-zero diagonal entry then A is aperiodic. ♡

Suppose that (X, σ) is a subshift of finite type. For $x, y \in X$, define $N(x, y) = \max\{i \mid x_j = y_j, |j| < i\}$. Given $\theta \in (0, 1)$, let d_θ be the metric on X defined by $d(x, y) = \theta^{N(x,y)}$, $(x, y \in X)$. The metric space (X, d_θ) is complete and defines the Tychonov topology on X.

Let $\mathcal{F}_\theta = \mathcal{F}_\theta(X)$ denote the space of continuous $\mathbb{C}$-valued Lipschitz functions on the metric space (X, d_θ). Given $f \in \mathcal{F}_\theta(X)$, let $|f|_\infty$ denote the supremum norm of f and $|f|_\theta$ denote the infimum over all Lipschitz constants C for f. If we define $\|f\|_\theta = |f|_\infty + |f|_\theta$, then $(\mathcal{F}_\theta(X), \|\,\|_\theta)$ is a Banach space. Let $\mathcal{F}_\theta(X)^\Gamma$ denote the closed subspace of $\mathcal{F}_\theta(X)$ consisting of Γ-invariant functions.

Given $A \in \mathbf{M}$, we let (X^+, σ) denote the associated one-sided subshift of finite type.

The previous results and definitions generalize in the obvious way to one-sided shifts. In particular, let d_θ^+ denote the metric on X^+ corresponding to d_θ and let $\mathcal{F}_\theta(X^+)$ denote the corresponding space of Hölder continuous functions on X^+. We recall there is a natural inclusion map $\mathcal{F}_\theta(X^+) \hookrightarrow \mathcal{F}_\theta(X)$.

Proposition 3.3.4 ([**58**, Proposition 1.2]). *Let $f \in \mathcal{F}_\theta(X)$. There exist $h \in \mathcal{F}_{\sqrt{\theta}}(X)$, $g \in \mathcal{F}_{\sqrt{\theta}}(X^+)$ such that $f = g + h - h \circ \sigma$. If $f \in \mathcal{F}_\theta(X)^\Gamma$, we may further require that both h and g are Γ-invariant.*

Proof. Only the last part is not in [**58**]. In order to obtain h, g Γ-invariant, we average the functions h, g defined in [**58**] over Γ. □

Let $f \in \mathcal{F}_\theta(X^+)$ be $\mathbb{R}$-valued. Let $L_f : \mathcal{F}_\theta(X^+) \to \mathcal{F}_\theta(X^+)$ denote the Ruelle operator defined by

$$L_f w(x) = \Sigma_{y \in \sigma^{-1}(x)}\, e^{f(y)} w(y), \ (w \in \mathcal{F}_\theta(X^+), x \in X).$$

We recall (part of) Ruelle's theorem, with (trivial) additions to take account of the presence of a Γ-action.

Theorem 3.3.5 (Ruelle [**58**, Theorem 2.2]). *Let $f \in \mathcal{F}_\theta(X^+)$ be real valued and $A \in \mathbf{M}^\Gamma$ be aperiodic. Then*

(a) *The operator L_f has a simple maximal positive eigenvalue β and corresponding strictly positive eigenfunction h. If f is Γ-invariant, so is h.*
(b) *The remainder of the spectrum of L_f is contained in a disk of radius smaller than β.*
(c) *There is a unique probability measure μ such that $L_f^\star \mu = \beta\mu$ (that is, $\int L_f v \, d\mu = \beta \int v \, d\mu$, all $v \in C^0(X^+)$). If f is Γ-invariant, so is μ.*

3.4. Equilibrium states

We continue to assume that (X, σ) is a subshift of finite type and A is aperiodic. Using the results of 3.3, one may show that associated to each $\mathbb{R}$-valued $f \in \mathcal{F}_\theta(X)$ there is a unique equilibrium state (measure) on X. We recall, without proof, the main definitions and results we need from [**9**, Chapter 4], [**58**, Chapter 3].

Fix $f \in \mathcal{F}_\theta(X)$. Let μ be a σ-invariant probability measure on X and denote the measure theoretic entropy of σ by $h_\mu(\sigma)$. The *pressure* of f is defined by

$$P(f) = \sup_\mu \{h_\mu(\sigma) + \int f \, d\mu\},$$

where the supremum is taken over all σ-invariant probability measures. We say that a σ-invariant probability measure m is an *equilibrium state* (for f) if $P(f) = h_m(\sigma) + \int f \, dm$.

THEOREM 3.4.1 ([**58**, Theorem 3.5]). *Every $f \in \mathcal{F}_\theta(X)$ has a unique equilibrium state m. Further,*

(a) *Relative to m, σ is Bernoulli (in particular, ergodic and strong mixing).*
(b) *$f, f' \in \mathcal{F}_\theta(X)$ have the same equilibrium state if and only if there exists $c \in \mathbb{R}$ such that f is cohomologous to $f' + c$.*
(c) *If f is cohomologous to $g \in \mathcal{F}_{\sqrt{\theta}}(X^+)$, then $P(f) = \log \beta$, where β is the maximal eigenvalue of the Ruelle operator L_g.*

If f is Γ-invariant, then the corresponding equilibrium measure is Γ-invariant.

Part 1

Markov partitions

CHAPTER 4

Markov partitions for finite group actions

4.1. Hyperbolicity

For the present we assume that Γ is finite. Let M be a (compact) riemannian Γ-manifold with associated Γ-invariant metric d on M. Let $f \in \mathrm{Diff}_\Gamma(M)$ and $\Lambda \subset M$ be compact, Γ- and f-invariant. Throughout this and subsequent sections, we make frequent use of the notational conventions established in Chapters 2 and 3.

Suppose that Λ is hyperbolic. Then there is a continuous f-invariant splitting $\mathbf{E}^s \oplus \mathbf{E}^u$ of $T_\Lambda M$ and $Tf|\mathbf{E}^s$, $Tf|\mathbf{E}^u$ satisfy the usual asymptotic estimates (we refer to Katok & Hasselblatt [**38**] for standard definitions and results on hyperbolicity). The next result follows easily using equivariance and uniqueness of the bundles $\mathbf{E}^s, \mathbf{E}^u$.

LEMMA 4.1.1. *The bundles $\mathbf{E}^s, \mathbf{E}^u$ are Γ-invariant subbundles of $T_\Lambda M$.*

Set $\mathbf{I} = \mathbf{I}(\Lambda)$. Since Γ is finite, so is $\mathbf{I}$. Given $H \in \mathbf{I}$, define

$$\mathbf{E}^{s,H} = T_{\Lambda^H} M^H \cap \mathbf{E}^s, \quad \mathbf{E}^{u,H} = T_{\Lambda^H} M^H \cap \mathbf{E}^u.$$

LEMMA 4.1.2. *For each $H \in \mathbf{I}$, $\mathbf{E}^{s,H}, \mathbf{E}^{u,H}$ are continuous $N(H)$-vector bundles over M^H. Further,*

(a) *$T_{\Lambda^H} M^H = \mathbf{E}^{s,H} \oplus \mathbf{E}^{u,H}$.*
(b) *Λ^H is a hyperbolic subset of $f|M^H$ with associated hyperbolic splitting $\mathbf{E}^{s,H} \oplus \mathbf{E}^{u,H}$.*
(c) *If Λ^H is not finite, then $\mathbf{E}^{s,H}$ and $\mathbf{E}^{u,H}$ are proper subbundles of $T_{\Lambda^H} M^H$.*

PROOF. Fix $H \in \mathbf{I}$. Since M^H, Λ^H are $N(H)$-invariant subsets of M, we may regard $\mathbf{E}^s, \mathbf{E}^u$ as $N(H)$-vector bundles over Λ^H. For $x \in \Lambda^H$, $T_x M$ has the structure of an H-representation and so we may write $T_x M = \mathbb{U} \oplus \mathbb{V}$, where $\mathbb{U}$ is the trivial factor of $T_x M$ and $\mathbb{V}$ is the sum of the non-trivial sub H-representations of $T_x M$. Since $\mathbb{U} = T_x M^H$, we see that $T_x M^H = \mathbf{E}^{s,H}_x \oplus \mathbf{E}^{u,H}_x$. Obviously, the asymptotic estimates drop to the splitting $\mathbf{E}^{s,H} \oplus \mathbf{E}^{u,H}$ and so Λ^H has the required hyperbolic structure. The final statement follows since $f|M^H \in \mathrm{Diff}(M^H)$. $\square$

Let $\delta > 0$ and $x \in \Lambda$. We define

$$W_\delta^s(x) = \{y \in M \mid d(f^n(x), f^n(y)) \le \delta,\, n \in \mathbb{N}\}. \tag{4.1}$$

We similarly define $W_\delta^u(x)$ using f^{-1}. It is well-known that for sufficiently small $\delta > 0$, independent of $x \in \Lambda$,

(1) $W_\delta^s(x), W_\delta^u(x)$ are smooth disks through x, tangent to $\mathbf{E}^s, \mathbf{E}^u$ at x,
(2) $W_\delta^s(x) \cap W_\delta^u(x) = \{x\}$.

Moreover, $W_\delta^s(x)$, $W_\delta^u(x)$ depend continuously on $x \in \Lambda$. We refer to $W_\delta^s(x)$, $W_\delta^u(x)$ as the local stable and unstable manifolds through the point x. In the sequel we often write $W_{\mathrm{loc}}^s(x)$, $W_{\mathrm{loc}}^u(x)$ and suppress reference to the constant δ. In any case, whether we write $W_\delta^{s/s}(x)$, or $W_{\mathrm{loc}}^{s/u}(x)$, we always assume that δ is chosen sufficiently small so that properties 1, 2 above hold.

Let $W^s(x)$, $W^u(x)$ denote the corresponding global stable and unstable manifolds. It follows from Γ-equivariance that for all $\gamma \in \Gamma$, $x \in \Lambda$ we have $W^s(\gamma x) = \gamma W^s(x)$, $W^u(\gamma x) = \gamma W^u(x)$. If $H \in \mathbf{I}$, $x \in \Lambda^H$, then $W^s(x) \cap M^H$ is the stable manifold at x of $f|M^H$. Similarly for the unstable manifold at x. If $x \in \Lambda_H$, then $W^s(x) \cap M^H \subset M_H$, similarly for the unstable manifold at x.

4.1.1. Local product structure.

DEFINITION 4.1.3 ([**38**, page 272]). The hyperbolic set Λ has *local product structure* if there exists $\delta = \delta_\Lambda > 0$ such that $W_\delta^s(x) \cap W_\delta^u(y) \subset \Lambda$ for all $x, y \in \Lambda$.

REMARKS 4.1.4. (1) It follows from the existence of local product structure that there exists $0 < a < \delta$ such that if $x, y \in \Lambda$ and $d(x, y) < a$ then $W_\delta^s(x) \cap W_\delta^u(y)$ consists of a single point $z = [x, y]$. Further, there exists $c > 1$ such that $d(x, z), d(z, y) \le c d(x, y)$ for all $x, y \in \Lambda$, $d(x, y) < a$. (2) If $f \in \mathrm{Diff}_\Gamma(M)$ and Λ is Γ-invariant, then $[\ ,\ ]$ is Hölder continuous and Γ-equivariant relative to the diagonal action of Γ on $\Lambda \times \Lambda$ [**38**, Chapter 19. §1]. (3) Local product structure is equivalent to local maximality [**38**, §18.4]. $\diamondsuit$

DEFINITION 4.1.5. Let f be a smooth equivariant diffeomorphism of M. A subset Λ of M is a Γ-*basic set* for f if

(a) Λ is a compact f- and Γ-invariant set.
(b) Λ is a hyperbolic set for f.
(c) Λ has local product structure.
(d) The induced map $\tilde{f} : \Lambda/\Gamma \to \Lambda/\Gamma$ is transitive.

REMARKS 4.1.6. (1) If Λ is Γ-basic, we have a finite decomposition $\Lambda = \cup_{i=1}^k \Lambda_i$ into topologically transitive components. Each Λ_i is a basic set for f, where the latter use of the term 'basic' is the conventional one, that is a topological Markov chain [**38**, Definition 6.4.18]. If we define $\Sigma_i = \{\gamma \in \Gamma \mid \gamma\Lambda_i = \Lambda_i\}$, then the Σ_i form a set of conjugate closed subgroups of Γ. (2) It follows from (1) and the transitivity of $\tilde{f}$ on Λ/Γ that an open dense subset of Λ consists of points with the *same* isotropy type. If particular, if Λ is a basic set (that is, $\Lambda = \Lambda_1$), then an open and dense subset of Λ consists of points with the same isotropy group (necessarily a normal subgroup of Γ). In either case, Λ has a unique minimal isotropy type. (3) Note that a Γ-invariant basic set is a Γ-basic set which is basic (that is, topologically transitive). Later, when we consider compact connected groups Γ, sets are typically connected, Γ-invariant and never hyperbolic. Consequently, there is little risk of confusion with the standard use of the term 'basic', and we shall drop the prefix 'Γ' and just refer to basic sets. $\diamondsuit$

4.2. Markov partitions & Equivariant symbolic dynamics

DEFINITION 4.2.1. Let Λ be a Γ-basic set. A closed set $R \subset \Lambda$ is a *proper rectangle* if $R = \overline{R^\circ}$ and $x, y \in R \Rightarrow [x, y] \in R$. ($R^\circ$ denotes the interior of R in Λ.)

If $x \in R \subset \Lambda$, we define

$$W^s(x, R) = W^s_\delta(x) \cap R, \quad W^u_\delta(x, R) = W^u_\delta(x) \cap R,$$

where we assume that $0 < \text{diameter}(R) \ll \delta$. We apply this definition when R is either a rectangle or the interior of a rectangle.

We recall the definition and properties of Markov partitions for basic sets. More details may be found in [**8, 9, 47**] or [**38**, §18.7].

DEFINITION 4.2.2. A finite set $\mathcal{R}$ of proper rectangles is a *Markov partition* for $f : \Lambda \to \Lambda$ if $\cup_{R\in\mathcal{R}} R = \Lambda$ and, for all $R, S \in \mathcal{R}$,

(a) $R^\circ \cap S^\circ = \emptyset$, if $R \neq S$.
(bs) $x \in R$, $f(x) \in S^\circ \Rightarrow f(W^s(x, R)) \subset W^s(f(x), S)$.
(bu) $x \in R$, $f^{-1}(x) \in S^\circ \Rightarrow f^{-1}(W^u(x, R)) \subset W^u(f^{-1}(x), S)$.

Let $\mathcal{R}$ be a Markov partition. Given $R \in \mathcal{R}$, define

$$\begin{aligned} \partial^s R &= \{x \in \partial R \mid W^s(x, R) \subset \partial R\}, \\ \partial^u R &= \{x \in \partial R \mid W^u(x, R) \subset \partial R\}. \end{aligned}$$

If we define $\partial^s \mathcal{R} = \cup_{R\in\mathcal{R}} \partial^s R$, and $\partial^u \mathcal{R} = \cup_{R\in\mathcal{R}} \partial^u R$, then

$$f(\partial^s \mathcal{R}) \subset \partial^s \mathcal{R}, \quad f^{-1}(\partial^u \mathcal{R}) \subset \partial^u \mathcal{R}. \tag{4.2}$$

Let Λ be a Γ-basic set and $\mathcal{R}$ be a Markov partition for $f : \Lambda \to \Lambda$. Define $\mathrm{mesh}(\mathcal{R}) = \max_{R \in \mathcal{R}} \mathrm{diameter}(R)$. For each $H \in \mathbf{I}(\Lambda)$, define

$$\mathcal{R}^H = \{R^H \mid R \in \mathcal{R} \text{ and } \overline{(R^\circ)^H} = R^H\}.$$

Without further conditions on $\mathcal{R}$, $\mathcal{R}^H$ may not be a Markov partition for $f : \Lambda^H \to \Lambda^H$. Indeed, $\mathcal{R}^H$ may be empty. Given $R \in \mathcal{R}$, let

$$\Gamma_R = \{\gamma \in \Gamma \mid \gamma R = R\}.$$

DEFINITION 4.2.3. A proper rectangle R is Γ-*admissible* if

(a) $\Gamma_R = \Gamma_{\bar{x}}$, for at least one $\bar{x} \in R^\circ$.
(b) If $R \cap \gamma R \neq \emptyset$, then $\gamma \in \Gamma_R$.

A Markov partition $\mathcal{R}$ for the Γ-equivariant map $f : \Lambda \to \Lambda$ is Γ-*admissible* if each $R \in \mathcal{R}$ is Γ-admissible.

DEFINITION 4.2.4. Let $\mathcal{R}$ be a Markov partition for $f : \Lambda \to \Lambda$. We say that $\mathcal{R}$ is a Γ-*invariant Markov partition* if

(a) $\mathcal{R}$ is Γ-admissible.
(b) Γ permutes the elements of $\mathcal{R}$.
(c) For all $H \in \mathbf{I}(\Lambda)$, $\mathcal{R}^H$ is a Markov partition for $f : \Lambda^H \to \Lambda^H$.

LEMMA 4.2.5. *Let $\mathcal{R}$ satisfy conditions (a,b) of Definition 4.2.4. In order that $\mathcal{R}$ be a Γ-invariant Markov partition, it suffices that for all $H \in \mathbf{I}(\Lambda)$, $\cup_{R \in \mathcal{R}^H} R^H = \Lambda^H$.*

PROOF. Fix $H \in \mathbf{I}(\Lambda)$ and let $R \in \mathcal{R}^H$. Since it follows from the Γ-equivariance of $[\ ,\ ]$ that R^H is a rectangle, it suffices to prove that $\overline{(R^H)^\circ} = R^H$. Since $(R^\circ)^H$ is an open subset of Λ^H, we have $(R^\circ)^H \subset (R^H)^\circ$, where the interior of R^H is taken in Λ^H. Taking closures, it follows that $\overline{(R^\circ)^H} \subset \overline{(R^H)^\circ}$. But $\overline{(R^\circ)^H} = R^H$ and so $R^H \subset \overline{(R^H)^\circ}$. Since the reverse inclusion is trivial, it follows that $\overline{(R^H)^\circ} = R^H$. □

THEOREM 4.2.6 ([**8, 9**]). *Let $f \in \mathrm{Diff}_\Gamma(M)$ and Λ be a Γ-basic set for f. Given $\varepsilon > 0$, there exists a Γ-invariant Markov partition $\mathcal{R}$ for $f : \Lambda \to \Lambda$ with mesh$(\mathcal{R}) < \varepsilon$.*

PROOF. The only statement not already in [**8, 9**] is the assertion that we can require the Markov partition to be Γ-invariant. In order to show this, we modify the construction given in [**8**] (see also the remark following the proof where we indicate an alternative construction based on shadowing). The idea is to start with a covering $\mathcal{M}$ of M by (interiors of) proper rectangles such that (a) Γ permutes the elements of $\mathcal{M}$, (b) each rectangle $R \in \mathcal{M}$ is Γ-admissible, and (c) $\overline{(R^\circ)^H} = R^H$, all $H \in \mathbf{I}(\Lambda)$, $R \in \mathcal{M}$. If these conditions hold, we say $\mathcal{M}$ is a Γ-cover. The naturality of the construction used in [**8**] implies that if we start with a Γ-cover, then the resulting Markov partition will be Γ-invariant (for condition (c) we use Lemma 4.1.2(c)). In what follows, we ignore the issue of the mesh of the partition. Indeed, exactly the same methods used in [**8**] (or [**9**]) show that the Markov partition can be constructed so as to have arbitrarily small mesh.

Let $x \in \Lambda$. We construct a proper Γ_x-invariant rectangle R containing x. Let A, B be Γ_x-invariant closed disk neighborhoods of x in $W^u_\varepsilon(x)$, $W^s_\varepsilon(x)$ respectively. We define $R = [A, B]$. Clearly, R is proper, Γ_x-invariant and $x \in R^\circ$. Moreover, if $x \in \Lambda^H$, then $R^H = [A^H, B^H]$ is also a proper rectangle (but now for $f|M^H$). All of this follows from the equivariance of the bracket $[\ ,\ ]$. We call a rectangle R constructed in this way a 'basic' rectangle.

We will construct a Γ-cover of Λ by basic rectangles. Let $\Lambda^1 \subset \ldots \subset \Lambda^N = \Lambda$ denote the filtration of Λ by submaximal isotropy type. Our construction proceeds by an upward induction on submaximal isotropy type. At stage i, we suppose that we have constructed a set $\mathcal{R}_i$ of basic rectangles for Λ such that

$$\text{(a) } \cup_{R\in\mathcal{R}_i} R^\circ \supset \Lambda^i;\ \text{(b) } \Gamma\text{-acts on } \mathcal{R}_i;$$
$$\text{(c) If } R \in \mathcal{R}_i \setminus \mathcal{R}_{i-1}, \text{ then } R \cap \Lambda^{i-1} = \emptyset.$$

Suppose $i = 1$. Let $\mathbf{I}_1 \subset \mathbf{I}(\Lambda)$ denote the set of maximal isotropy groups. Then $\Lambda^1 = \cup_{H\in\mathbf{I}_1}\Lambda^H$ and each Λ^H is a compact subset of Λ. Let $\tilde{\mathbf{I}}_1 = \{H_1, \ldots, H_k\}$ be a set of representatives for the conjugacy classes of the groups in $\mathbf{I}_1$. Let $\tilde{\Lambda}^1 = \cup_{i=1}^k \Lambda^{H_i}$. Since $\tilde{\Lambda}^1$ is compact, we can choose a finite set of basic rectangles, say $\tilde{\mathcal{R}}_1$, such that the interiors of the rectangles in $\tilde{\mathcal{R}}_1$ cover $\tilde{\Lambda}^1$, the interior of every rectangle meets $\tilde{\Lambda}^1$ and, if $R \in \tilde{\mathcal{R}}_1$, then $R \cap \Lambda^H \neq \emptyset$ if and only if $\Gamma_R = H$. Let $\mathcal{R}_1 = \{\gamma R \mid \gamma \in \Gamma,\ R \in \tilde{\mathcal{R}}_1\}$. It follows from our construction that $\mathcal{R}_1$ satisfies (a,b). Suppose next that we have constructed $\mathcal{R}_{i-1}$ satisfying (a,b,c), $1 < i < N$. Observe that the complement in Λ^i of the union of the interiors of the rectangles in $\mathcal{R}_{i-1}$ is a compact Γ-invariant set, say $\hat{\Lambda}^i$. We now just repeat the construction we gave for $i = 1$ but now applied to $\hat{\Lambda}^i$. Since the cover $\mathcal{R}_N$ of Λ is obviously a Γ-cover, Bowen's construction carries through to yield the required Γ-invariant Markov partition for Λ. □

REMARK 4.2.7. An alternative construction of Γ-invariant Markov partitions can be based on shadowing [**9**]. For sufficiently small $\gamma > 0$ (we follow the notation and terminology of [**66**, 9.6]), we construct a finite γ-dense Γ-invariant subset of Λ which induces γ-dense Γ-invariant subsets of all the orbit strata. The construction is the obvious upward induction on orbit strata and the rest of proof follows [**66**]. ♢

4.2.1. Symbolic dynamics for Γ-basic sets. Let $f \in \mathrm{Diff}_\Gamma(M)$, Λ be a Γ-basic set for f and $\mathcal{R}$ be a Γ-invariant Markov partition for Λ. Suppose that $\mathcal{R}$ contains n rectangles which we label $R_1, \ldots, R_n$. Let A denote the associated $n \times n$ 0-1 matrix and $\sigma : \Sigma_A = \Sigma \to \Sigma$ denote the resulting subshift of finite type. Since Γ acts on $\mathcal{R}$, it follows that Σ is a Γ-subshift of finite type.

More generally, let $\Lambda^1 \subset \ldots \subset \Lambda^N = \Lambda$ be the filtration of Λ by submaximal isotropy type. Since $\mathcal{R}$ is Γ-invariant, $\mathcal{R}$ determines a Γ-invariant Markov partition $\mathcal{R}^j$ for $f : \Lambda^j \to \Lambda^j$, $1 \leq j \leq N$. Hence, we may define an associated subshift $\sigma : \Sigma^j \to \Sigma^j$, $1 \leq j \leq N$. Of course, $\Sigma^N = \Sigma$. For future reference, note that if $\tau \in \mathcal{M}_\ell(\Lambda)$ then

$$\Sigma_H = \Sigma^k_H, \quad k \geq \ell, \; H \in \tau. \tag{4.3}$$

THEOREM 4.2.8 ([**9**],[**58**, Appendix III]). *The map* $\pi : \Sigma \to \Lambda$ *defined by*

$$\pi(x) = \cap_{k=-\infty}^{\infty} f^{-k}(R_{x_k})$$

is well-defined. Furthermore,

(a) *π is Hölder continuous and surjective.*
(b) *π is 1:1 on a residual subset of Σ.*
(c) *$\#\pi^{-1}(x) \leq n^2$ for all $x \in \Lambda$.*
(d) *$f\pi = \pi\sigma$.*
(e) *π is Γ-equivariant.*
(f) *Every isotropy group in Λ occurs as the isotropy group of a point in Σ.*

PROOF. Everything except (e,f) is proved in [**8, 9**]. Statement (e) is immediate from the fact that Γ acts on $\mathcal{R}$. It remains to prove (f). Suppose $H \in \mathbf{I}(\Lambda)$. Since $\mathcal{R}$ is a Γ-invariant Markov partition, it follows that $\mathcal{R}^H$ is a Markov partition for $f : \Lambda^H \to \Lambda^H$. Now use (b) and the fact that if $\Lambda_H \neq \emptyset$, then Λ^H contains interior points with isotropy H. □

THEOREM 4.2.9. *Let Λ be a Γ-basic set for f and $\pi : \Sigma \to \Lambda$ be as in Theorem 4.2.8. Then π preserves isotropy type. That is, for all $\tau \in \mathcal{O}(\Lambda)$, we have*

$$\pi^{-1}(\Lambda_\tau) = \Sigma_\tau.$$

In particular, $\mathbf{I}(\Sigma) = \mathbf{I}(\Lambda)$.

PROOF. It suffices to show that if $H \in \mathbf{I}(\Lambda)$, then $\pi^{-1}(\Lambda^H) = \Sigma^H$. We prove this by an upward induction over the submaximal isotropy type of $H \in \mathbf{I}(\Lambda)$. We present the first stage of the induction, leaving the general step to the reader. Suppose then that $H \in \mathbf{I}(\Lambda)$ is maximal. It follows by Γ-equivariance of π that $\pi^{-1}(\Lambda^H) \supset \Sigma^H$. Since $\mathcal{R}$ is Γ-invariant and H is maximal, if $R \in \mathcal{R}$ and $R \cap \Lambda^H \neq \emptyset$, then $\Gamma_R = H$. Consequently, if $x = (x_i) \in \pi^{-1}(\Lambda^H)$, then $\Gamma_{R_{x_i}} = H$, all $i \in \mathbb{Z}$ (otherwise $\pi(x) \notin \Lambda^H$). Hence $\pi^{-1}(\Lambda^H) = \Sigma^H$. □

EXAMPLES 4.2.10. (1) Suppose $f \in \mathrm{Diff}_\Gamma(M)$ and Λ is a Γ-basic set for f. As an immediate consequence of our results, there exist a subgroup H of Γ and compact $\Xi \subset \Lambda$ such that

(1) Ξ is H-invariant and $\Gamma(\Xi) = \Lambda$.
(2) Ξ is a basic set for f. In particular, $f|\Xi$ has a dense orbit.
(3) If $\phi : \Lambda \to \mathbb{R}$ is Hölder and Γ-invariant then there is a unique Γ-invariant measure μ_ϕ on Λ such that $m = \mu_\phi|\Xi$ is the equilibrium measure of $\phi|\Xi$. Further, $\{m\gamma \mid \gamma \in \Gamma\}$ is the ergodic decomposition of μ_ϕ.

(2) Suppose that Λ is a basic set for $f \in \mathrm{Diff}_\Gamma(M)$. Suppose that Γ contains an involution η such that $M^\eta = \mathrm{Fix}(\eta)$ is connected and separates M into two connected components. Then

$$\Lambda \cap M^\eta \neq \emptyset \Longrightarrow \Lambda \subset M^\eta.$$

Suppose $\Lambda \not\subset M^\eta$. Taking the spectral decomposition of $f|\Lambda$, it is no loss of generality to assume that $f|\Lambda$ is topologically mixing. Now we derive a contradiction by choosing open subsets U, V of Λ which lie in different connected components of $M \setminus M^\eta$. ♡

PROPOSITION 4.2.11. *Let Λ be a Γ-basic set for f and $\pi : X = \Sigma \to \Lambda$ be as in Theorem 4.2.8. For each non-minimal $H \in \mathbf{I}(\Lambda)$, we can find an open neighborhood U_H of Λ^H in Λ such that if $x \in U^H \setminus \Lambda^H$, there exists $n \in \mathbb{Z}$ such that $f^n(x) \notin U^H$.*

PROOF. The result follows from Theorem 4.2.9. □

We recall that $f : \Lambda \to \Lambda$ is *expansive* if there exists $\varepsilon > 0$ such that if x, y are distinct points of M, with $x \in \Lambda$, then there exists $n \in \mathbb{Z}$ such that $d(f^n(x), f^n(y)) > \varepsilon$. If f is Γ-equivariant, then it is natural to say that f is "Γ-expansive" if there exists $\varepsilon > 0$ such that if α, β are distinct Γ-orbits, then there exists $n \in \mathbb{Z}$ such that $d(f^n(\alpha), f^n(\beta)) \geq \varepsilon$. However, if the Γ-action is not free then f is generally not Γ-expansive

EXAMPLE 4.2.12. Let $\mathbb{Z}_2(\kappa)$ act on $\mathbf{3} = \{-1, 0, 1\}$ by $\kappa(-1) = 1$, $\kappa(0) = 0$. Let $\Lambda = \mathbf{3}^{\mathbb{Z}}$ denote the corresponding $\mathbb{Z}_2$-equivariant full shift on three symbols. We can realize Λ as the basic set of a $\mathbb{Z}_2$-equivariant diffeomorphism $f : \mathbb{R}^2 \to \mathbb{R}^2$, where $\mathbb{Z}_2$ acts on $\mathbb{R}^2$ as minus the identity map (see [**22**, §2, Example]). Let I^s, I^u denote open intervals, centered at zero, contained in the y- and x-axes respectively. We may suppose f is chosen so that $W^s_\varepsilon(0) = I^s$, $W^u_\varepsilon(0) = I^u$ and f is linear on $I^u \times I^s$. Let $\rho > 0$. We shall find a pair of $\mathbb{Z}_2$-orbits whose iterates under f are always of distance less than ρ apart. Let $y \in I^s$, $x \in I^u$, where $x, y > 0$ and $(x, \pm y) \in \Lambda$. Then $(x, -y) \in W^s(x, y)$ and $(-x, y) \in W^u(x, y)$. Choose $N > 0$ so that $d(f^N(0, y), d^N(0, -y)) < \rho$. Then choose x small so that $f^n(\pm x, y) \in I^u \times I^s$, and $d(f^n(x, y), f^n(-x, y)) < \rho$, $0 \leq n \leq N$. But then $d(f^n(x, y), f^n(x, -y)) < \rho$, $n \geq N$. Since $d(f^n(x, y), f^n(-x, y)) < \rho$, $n \leq N$, it follows that the distance between the $\mathbb{Z}_2$-orbits of (x, y) and $(x, -y)$ is always less than ρ. ♡

REMARKS 4.2.13. (1) It follows from example 4.2.12 that the dynamics on the orbit space Λ/Γ will not usually be expansive. This has been observed previously [**18**]. Later, we consider to what extent shadowing properties hold on Λ/Γ. (2) If dynamics on the orbit space is not expansive and Λ is zero dimensional (equivalently, a subshift of finite type), then dynamics on Λ/Γ *cannot* be topologically conjugate to a subshift of finite type [**47**, Proposition 11.9] ◇

Using Theorem 4.2.8, we may extend a number of well-known results from ergodic theory to the equivariant context. As an example, we cite

THEOREM 4.2.14 ([**9**]). *Let Λ be a Γ-invariant hyperbolic basic set for the equivariant diffeomorphism f. Suppose that $\phi : \Lambda \to \mathbb{R}$ is Hölder and Γ-invariant. Then ϕ has a unique equilibrium state μ_ϕ, μ_ϕ is Γ-invariant, ergodic and even Bernoulli if $f|\Lambda$ is topologically mixing.*

4.2.2. Markov partitions on Λ/Γ. Although every Γ-invariant Markov partition on Λ determines an equivariant symbolic dynamics on Λ, we cannot generally use the coding of $f : \Lambda \to \Lambda$ to determine a coding of $\tilde{f} : \Lambda/\Gamma \to \Lambda/\Gamma$. In order to do this, we require new conditions on Markov partitions.

DEFINITION 4.2.15. Let Λ be a Γ-basic set. A Markov partition $\mathcal{R}$ for Λ is Γ-*regular* if

(a) Γ acts freely on $\mathcal{R}$.
(b) If $R, S \in \mathcal{R}$ and $f(R) \cap S^\circ \neq \emptyset$ then $f(R) \cap \gamma S^\circ = \emptyset$ for all $\gamma \neq e$.
(c) For all $\gamma \neq e$, $R \in \mathcal{R}$, $R \cap \gamma R \subset \Lambda_S$.

REMARK 4.2.16. Notice that it follows from (a) that $R \cap \gamma R$ is contained in $\partial R \cap \partial(\gamma R)$ for $\gamma \neq e$. Combined with (c), this implies $R \cap \gamma R \subset \partial R \cap \partial(\gamma R) \cap \Lambda_S$. ◇

The proof of the following lemma is a simple consequence of (b).

LEMMA 4.2.17. *Suppose that $\mathcal{R}$ is a Γ-regular Markov partition. Let A denote the associated 1-matrix. Then if $a_{ij} = 1$, $a_{i\gamma(j)} = 0$, for all $\gamma \neq e$.*

In the sequel, we refer to a 01-matrix possessing this property as Γ-regular.

We adopt some notational conventions that prove useful in the sequel. If $X \subset \Lambda$, we set $p(X) = \tilde{X}$. Thus, $\tilde{\Lambda} = \Lambda/\Gamma$. If $\tilde{X}$ is a closed subset of $\tilde{\Lambda}$, we define $\tilde{X}^\sigma = \tilde{X} \setminus (\tilde{\Lambda}_S \cap \partial \tilde{X})$.

Associated to a Γ-regular Markov partition $\mathcal{R}$ of Λ we define $\tilde{\mathcal{R}} = \{\tilde{R} = p(R) \mid R \in \mathcal{R}\}$. Obviously $\tilde{\mathcal{R}}$ is a cover of Λ/Γ and, since Γ acts on $\mathcal{R}$, $\tilde{R}^\circ \cap \tilde{S}^\circ = \emptyset$ unless $R = \gamma S$, some $\gamma \in \Gamma$. Note that $\tilde{R} \setminus \tilde{R}^\sigma \subset \partial \tilde{R}$.

LEMMA 4.2.18. *Let $\tilde{R} \in \tilde{\mathcal{R}}$. For each $\alpha \in \tilde{R}^\sigma$, the sets $W^s(\alpha, \tilde{R}) = pW^s(x, R)$, $W^u(\alpha, \tilde{R}) = pW^u(x, R)$, are defined independently of the choice of $R \in \mathcal{R}$, such that $p(R) = \tilde{R}$, and $x \in R$ such that $p(x) = \alpha$.*

PROOF. It follows from (c) that, once R is chosen, $R \cap \alpha$ consists of a single point. The remainder of the statement follows trivially from equivariance. □

The same argument also proves

LEMMA 4.2.19. *Let $\mathcal{R}$ be Γ-regular. For each $\tilde{R} \in \tilde{\mathcal{R}}$, we have a well-defined continuous map*

$$[\ ,\ ] : \tilde{R}^\sigma \times \tilde{R}^\sigma \to \tilde{R},$$

characterized by $[\alpha, \beta] = W^s(\alpha, \tilde{R}) \cap W^u(\beta, \tilde{R})$. In terms of the bracket on Λ, $[\alpha, \beta] = [x, y]$, where $x \in R \cap \alpha$, $y \in R \cap \beta$ and $p(R) = \tilde{R}$.

Since $\mathcal{R}$ is a Markov partition on Λ, it follows from the previous two lemmas that $\tilde{\mathcal{R}}$ is a Markov partition on Λ/Γ. Specifically,

PROPOSITION 4.2.20. *If $\mathcal{R}$ is Γ-regular, then*

(1) $\cup_{\tilde{R} \in \tilde{\mathcal{R}}} \tilde{R} = \Lambda/\Gamma$.

(2) $\overline{\tilde{R}^\circ} = \tilde{R}$, *all* $\tilde{R} \in \tilde{\mathcal{R}}$.

(bs) $\alpha \in \tilde{R}$, $\tilde{f}(\alpha) \in \tilde{S}^\circ \Rightarrow \tilde{f}(W^s(\alpha, \tilde{R})) \subset W^s(\tilde{f}(\alpha), \tilde{S})$.

(bu) $\alpha \in \tilde{R}$, $\tilde{f}^{-1}(\alpha) \in \tilde{S}^\circ \Rightarrow \tilde{f}^{-1}(W^u(\alpha, \tilde{R})) \subset W^u(\tilde{f}^{-1}(\alpha), \tilde{S})$.

Suppose that $\mathcal{R}$ is a Γ-regular Markov partition for the Γ-basic set Λ. Let A denote the associated 01 matrix A, and $\Sigma_A \subset \mathbf{n}^{\mathbb{Z}}$ the corresponding subshift of finite type, where n is the number of rectangles in $\mathcal{R}$. The next proposition uses condition (a) of our definition of Γ-regular Markov partition.

PROPOSITION 4.2.21. *Let $\mathcal{R}$ be a Γ-regular Markov partition for the Γ-basic set Λ. Denote the associated subshift of finite type by $\Sigma_A \subset \mathbf{n}^{\mathbb{Z}}$. There is a natural free action of Γ on Σ_A and the projection map $\pi : \Sigma_A \to \Lambda$ is Γ-equivariant.*

REMARK 4.2.22. Unlike what happens for Γ-invariant Markov partitions, π does *not* preserve isotropy type (unless the action of Γ on Λ is free). ♢

LEMMA 4.2.23. *Suppose that $\Sigma_A \subset \mathbf{n}^{\mathbb{Z}}$ is the Γ-subshift of finite type defined by the Γ-regular 01-matrix A. Let $\rho : \mathbf{n} \to \tilde{\mathbf{n}}$ denote the corresponding orbit map (thus $\tilde{n} = n/|\Gamma|$). Then*

(a) *If we let $\tilde{A}$ be the $\tilde{n} \times \tilde{n}$ 01 matrix defined by the condition $\tilde{A}(i,j) = 1$ if and only if there exist $I, J \in \mathbf{n}$ such that $\rho(I) = i$, $\rho(J) = j$, and $A(I,J) = 1$, then $\Sigma_A/\Gamma \approx \Sigma_{\tilde{A}}$.*

(b) *The orbit map $p : \Sigma_A \to \Sigma_A/\Gamma$ is given by*

$$p((x_i)) = (\rho(x_i)), \quad ((x_i) \in \Sigma_A).$$

PROOF. A trivial exercise (cf Example 3.2.4). □

PROPOSITION 4.2.24. *Let $\mathcal{R} = \{R_i \mid i \in I\}$ be a Γ-regular Markov partition for the Γ-basic set Λ. Denote the associated subshift of finite type by $\Sigma_A \subset \mathbf{n}^{\mathbb{Z}}$. Let $\tilde{f}$ denote the induced homeomorphism induced by f on Λ/Γ. Set $\tilde{\mathcal{R}} = \{\tilde{R}_i = p(R_i) \mid i \in I\}$. Define $\tilde{\pi} : \Sigma_{\tilde{A}} \to \Lambda/\Gamma$ by*

$$\tilde{\pi}((\tilde{x}_i)) = \cap_{i\in\mathbb{Z}} \tilde{f}^{-i}(\tilde{R}_i).$$

Then

(a) *$\tilde{\pi} : \Sigma_{\tilde{A}} \to \Lambda/\Gamma$ is a well defined Hölder continuous surjection. In particular, $\tilde{\pi}(x)$ consists of a single point for all $x \in \Sigma_{\tilde{A}}$.*

(b) *$\tilde{\pi} \circ p = p \circ \pi$.*

PROOF. Let $(\tilde{x}_i) \in \Sigma_{\tilde{A}}$. Pick $(x_i) \in \Sigma_A$ such that $p((x_i)) = (\tilde{x}_i)$. Define $z = p\pi((x_i))$. Obviously, z is independent of the choice of (x_i). Further, since $\pi((x_i)) = \cap_{i\in\mathbb{Z}} f^{-i}(R_i)$, it follows that $(\tilde{x}_i) = \cap_{i\in\mathbb{Z}} \tilde{f}^{-i}(\tilde{R}_i)$. Hence, $\tilde{\pi}$ is a well-defined as a map and that $\tilde{\pi} \circ p = p \circ \pi$. Since π and the orbit map p are surjective it follows that $\tilde{\pi}$ is surjective. It remains to prove that $\tilde{\pi}$ is Hölder continuous. Continuity is immediate. Since the action of Γ on Σ_A is free and $\Sigma_{\tilde{A}}$ is totally disconnected, it follows that the quotient $p : \Sigma_A \to \Sigma_{\tilde{A}}$ admits a global Hölder continuous section ξ. But then $\tilde{\pi} = p \circ \pi\xi$ is a composite of Hölder continuous maps. □

REMARK 4.2.25. Note that the conclusion of the Proposition holds under the assumption that $\mathcal{R}$ satisfies conditions (a,b). Of course, if (c)

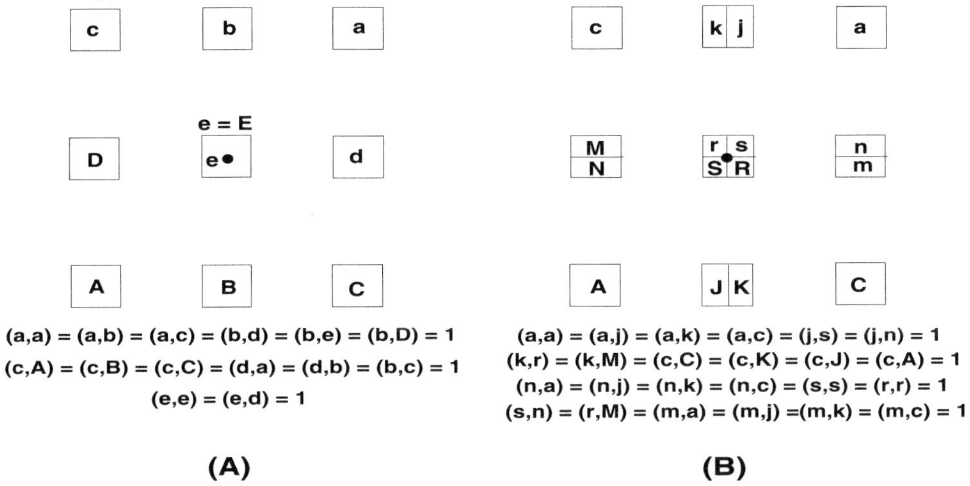

FIGURE 1. Γ-invariant & Γ-regular Markov partitions

fails, then we will not generally be able to define a bracket operation on the sets $\tilde{R}^\sigma$. ♢

EXAMPLE 4.2.26. We continue with the assumptions of Example 4.2.12. In Figure 1 (A), we show a Γ-invariant Markov partition for $\Lambda \subset \mathbb{R}^2$. We have labeled the proper rectangles $a, \ldots, A$, where X is the symmetric image of the rectangle x. We display a generating set of relations for the 01 matrix. Note that, by Γ-equivariance, $(x, y) = 1$ if and only if $(X, Y) = 1$. This partition is not Γ-regular. For example, we have $(b, d) = (b, D) = 1$. In Figure 1 (B), we show a refined Γ-invariant Markov partition for Λ which is Γ-regular. Note that we could replace M, N and m, n by the two rectangles $M \cup N$ and $m \cup n$ and still have Γ-regularity. ♡

4.3. Examples of symmetric hyperbolic basic sets: Γ finite

We describe two classes of examples of symmetric hyperbolic sets. The first class comprises an equivariant version of the Smale horseshoe, the second a class of equivariant solenoids.

4.3.1. Equivariant horseshoes. We follow the notation of section 3.2. Our first result [**22**], shows that equivariant subshifts of finite type can be realized as indecomposable pieces of the omega set of equivariant Axiom A diffeomorphisms.

THEOREM 4.3.1 ([**22**, §7]). *Let $A \in \mathbf{M}(n)^\Gamma$ and (X_A, σ) be the associated subshift of finite type. Then we may construct a compact*

Γ-manifold M and smooth Γ-equivariant diffeomorphism $f : M \to M$ such that

(a) *f satisfies Axiom A.*
(b) *There is a compact Γ- and f-invariant hyperbolic subset Z of $\Omega(f)$ such that Z is a union of pieces from the spectral decomposition of $\Omega(f)$ and $f|Z : Z \to Z$ is Γ-equivariantly conjugate to (X_A, σ).*

REMARK 4.3.2. We may take M to be the unit sphere in an appropriate representation of Γ. ♢

4.3.2. Equivariant attractors. We review the results of Field et al [**28**] on the existence of symmetric hyperbolic attractors. We emphasize the case of diffeomorphisms, though the results have analogues for flows.

Let Λ be a compact subset of the Γ-manifold M. The symmetry group of Λ is the subgroup Γ_Λ of Γ leaving Λ invariant.

THEOREM 4.3.3 ([**28**, Theorem 4.7]). *Suppose that M is a connected Γ-manifold, dim$(M) \geq 4$. Let M_0 denote the set of points with trivial isotropy type and assume that $M_0 \neq \emptyset$. Let H be a subgroup of Γ which fixes a connected component of M_0. There exists a smooth Γ-equivariant diffeomorphism f of M such that f has a connected hyperbolic basic attractor $A \subset M_0$ with $\Gamma_A = H$. An analogous result for flows holds if dim$(M) \geq 5$.*

REMARK 4.3.4. Using the methods of [**28**, §6], it is possible to construct hyperbolic attractors supported on orbit strata other than the principal stratum. Although this is not explicitly done in [**28**, §6], it is also possible to construct attractors which contain points of more than one isotropy type. We give a simple example below. We refer to [**31**] for an example of a non-uniformly hyperbolic $\mathbb{Z}_2$-invariant attractor. ♢

EXAMPLE 4.3.5. We recall that the classical solenoid $\mathcal{S}_2$ is defined as the inverse limit of the expanding map $f : S^1 \to S^1$, $f(\theta) = 2\theta$ [**75**]. Let $\mathbb{Z}_2$ act on $\mathbb{R}^3$ as $(x, y, z) \mapsto (x, -y, -z)$. We show that $\mathcal{S}_2$ can be realized as a hyperbolic $\mathbb{Z}_2$-invariant attractor of a $\mathbb{Z}_2$-equivariant diffeomorphism of $\mathbb{R}^3$ and that the action of $\mathbb{Z}_2$ on $\mathcal{S}_2$ is not free.

The map f is obviously equivariant with respect to the $\mathbb{Z}_2$-action on S^1 defined by $\theta \mapsto -\theta$. The $\mathbb{Z}_2$-action lifts to the action on $\mathcal{S}_2$ defined by $(\theta_0, \theta_1, \ldots) \mapsto (-\theta_0, -\theta_1, \ldots)$. The fixed point set of this action is $\{0^p\pi^\infty \mid p \geq 0\} \cup \{0^\infty\}$ (0^p denotes 0 repeated p times).

Let $D = \{z \in \mathbb{C} \mid |z| \leq 1\}$ and set $\mathbf{T} = D \times S^1$ (the solid torus). We define a $\mathbb{Z}_2$ action on $\mathbf{T}$ by $(z, \theta) \mapsto (\bar{z}, -\theta)$ and note that $\mathbf{T}$

equivariantly embeds in $\mathbb{R}^3$ – the 'core' of the torus map be mapped to the unit circle in the (x,y)-plane. If we define $F : \mathbf{T} \to \mathbf{T}$ by $F(z,\theta) = (\frac{z}{4} + \frac{1}{2}\exp(\imath\theta), 2\theta)$, then F is a $\mathbb{Z}_2$-equivariant embedding. Let $\Lambda = \cap_{n\geq 0} F^n(\mathbf{T})$. The compact set Λ is a hyperbolic F- and $\mathbb{Z}_2$-invariant subset of $\mathbf{T}$. There is a $\mathbb{Z}_2$-equivariant homeomorphism $\chi : \mathcal{S}_2 \to \Lambda$ such that $F\chi = \chi\hat{f}$, where $\hat{f} : \mathcal{S}_2 \to \mathcal{S}_2$ is the homeomorphism of $\mathcal{S}_2$ induced by f. Finally, the map $F : \mathbf{T} \to \mathbf{T}$ is $\mathbb{Z}_2$-equivariantly isotopic in $\mathbb{R}^3$ to the identity map of $\mathbf{T}$ and so extends by the equivariant isotopy theorem to a smooth $\mathbb{Z}_2$-equivariant diffeomorphism of $\mathbb{R}^3$ (we refer to [**29**] for more details on this construction). ♡

4.4. Existence of Γ-regular Markov partitions

Our aim in this section is to prove the following basic result.

THEOREM 4.4.1. *Let $f : \Lambda \to \Lambda$ be a Γ-basic set, Γ-finite. Let $\varepsilon > 0$ and suppose that the subset $\Lambda_\Pi \subset \Lambda$ of points with trivial isotropy is open and dense in Λ. Then Λ admits a Γ-regular Markov partition $\mathcal{M}$ with mesh($\mathcal{R}$) $< \varepsilon$.*

PROOF. Our proof of Theorem 4.4.1 depends on constructing a refinement of a Γ-invariant Markov partition of Λ.

It follows from Theorem 4.2.6 that we may choose a Γ-invariant Markov partition $\mathcal{R}$ of Λ such that $\text{mesh}(\mathcal{R}) < \varepsilon$. Let $\Lambda_S = \Lambda \cap M_S$ denote the set of singular Γ-orbits in Λ. Define

$$\mathcal{R}^1 = \{R \in \mathcal{R} \mid R \cap \Lambda_S \neq \emptyset\}, \quad \mathcal{R}^2 = \mathcal{R} \setminus \mathcal{R}^1.$$

It follows from Theorem 4.2.9 that $\mathcal{R}^2 \neq \emptyset$.

Since $\mathcal{R}$ is a Γ-invariant Markov partition, Γ acts freely on $\mathcal{R}^2$ and $\gamma R \cap R = \emptyset$, $\gamma \neq e$, $R \in \mathcal{R}^2$. Refining rectangles in $\mathcal{R}^2$ if necessary, we can always assume that if $S, T \in \mathcal{R}^2$ and $f(S^\circ) \cap R^\circ \neq \emptyset$ then $f(S^\circ) \cap \gamma R^\circ = \emptyset$, $\gamma \neq e$.

Let $\mathbf{R}^1 = \cup_{R\in\mathcal{R}^1} R$. Set $\mathbf{U} = \mathbf{R}^1 \setminus \cup_{x\in\Lambda_S}(W^s_{\text{loc}}(x) \cup W^u_{\text{loc}}(x))$. Using either the symbolic dynamics provided by $\mathcal{R}$ or the local product structure, it is easy to verify that $\mathbf{U}$ is an open and dense Γ-invariant subset of $\mathbf{R}^1$. Provided $\varepsilon > 0$ is sufficiently small, given any $x \in \mathbf{U}$, there exist smallest $N = N(x), M = M(x) > 0$ such that $f^i(x) \in \mathbf{U}$, $-M < i < N$, and $f^N(x), f^{-M}(x) \notin \mathbf{R}^1$.

Fix $R \in \mathcal{R}^1$, $S, T \in \mathcal{R}^2$. For $m, n \in \mathbb{N}$, define the following open subsets of R°:

$$\begin{aligned} R(S)^\circ_m &= \{x \in R^\circ \cap \mathbf{U} \mid m = M(x),\ f^{-M(x)}(x) \in S^\circ\}, \\ R(T)^\circ_n &= \{x \in R^\circ \cap \mathbf{U} \mid n = N(x),\ f^{N(x)}(x) \in T^\circ\}, \\ R(S,T)^\circ_{m,n} &= R(S)^\circ_m \cap R(T)^\circ_n. \end{aligned}$$

Set

$$\begin{aligned} R(S)^{\circ\circ} &= \cup_{m\geq 1} R(S)^{\circ}_{m},\ R(T)^{\circ\circ} = \cup_{n\geq 1} R(T)^{\circ}_{n}, \\ R(S,T)^{\circ\circ} &= R(S)^{\circ\circ} \cap R(T)^{\circ\circ}, \\ &= \cup_{m,n\geq 1} R(S,T)^{\circ}_{m,n}, \\ &= \{x \in R^{\circ} \cap \mathbf{U} \mid f^{-M(x)}(x) \in S^{\circ}, f^{N(x)}(x) \in T^{\circ}\}. \end{aligned}$$

Let $R(S)$ denote the closure (in Λ) of $R(S)^{\circ\circ}$. We similarly define $R(T)$ and $R(S,T)$. Note that $R(S,T)$ contains $R(S,T)^{\circ\circ}$ as an open and dense set and that, in general, $R(S,T)^{\circ} \supsetneq R(S,T)^{\circ\circ}$.

LEMMA 4.4.2. *Let $R \in \mathcal{R}^1$, $S,T \in \mathcal{R}^2$. If $x \in R^{\circ}$, $f^{-M(x)} \in S^{\circ}$, $f^N(x) \in T^{\circ}$, then for each i, $-M < i < N$, there exists a unique $P_i \in \mathcal{R}^1$ such that $f^i(x) \in P_i^{\circ}$.*

PROOF. Suppose there exists i, $-M < i < N$, such that $f^i(x) \in \partial P$. Since $\partial P = \partial^s P \cup \partial^u P$, it follows that either $f^i(x) \in \partial^s P$ or $f^i(x) \in \partial^u P$. Suppose that $i > 0$ and $f^i(x) \in \partial^u P$. It follows from (4.2) that $x \in \partial^u R$, contradicting our assumption that $x \in R^{\circ}$. The other cases are similarly excluded. □

LEMMA 4.4.3. *Let $R \in \mathcal{R}^1$, $S,T \in \mathcal{R}^2$. If $x \in R^{\circ}$, $f^{-M}(x) \in S^{\circ}$, $f^N(x) \in T^{\circ}$ then*

$$W^s(x, R^{\circ}) \subset R(T)^{\circ}, \tag{4.4}$$

$$W^u(x, R^{\circ}) \subset R(S)^{\circ}, \tag{4.5}$$

$$f^N(W^s(x, R(T)) \subset W^s(f^N(x), T), \tag{4.6}$$

$$f^{-M}(W^u(x, R(S))) \subset W^u(f^{-M}(x), S). \tag{4.7}$$

PROOF. It follows from Lemma 4.4.2, that for each $i \in \{1, N-1\}$, there exists a unique $P_i \in \mathcal{R}^1$ such that $f^i(x) \in P_i^{\circ}$. But then

$$f(W^s(f^i(x), T_i^{\circ})) \subset W^s(f^{i+1}(x), T_{i+1}^{\circ}).$$

Hence $f^N(W^s(x, R^{\circ}) \subset W^s(f^N(x), T^{\circ})$, proving (4.4). The proofs of the remaining statements are similar and omitted. □

LEMMA 4.4.4. *Let $R \in \mathcal{R}^1$, $S,T \in \mathcal{R}^2$ and suppose $R(S,T) \neq \emptyset$. Then $R(S,T)$ is a rectangle.*

PROOF. Since $[\ ,\ ]$ is continuous, it suffices to prove that for all $x, y \in R(S,T)^{\circ\circ}$, $[x,y] \in R(S,T)$. Suppose then that $x \in R(S,T)^{\circ}_{m,n}$, $y \in R(S,T)^{\circ}_{p,q}$. Necessarily, $x \in R(T)^{\circ}_{n}$, $y \in R(S)^{\circ}_{p}$. Hence, by Lemma 4.4.3, $[x,y] \in R(S,T)^{\circ}_{p,n} \subset R(S,T)$. Finally, since $R(S,T)^{\circ\circ}$ is an open and dense subset of $R(S,T)^{\circ}$, it is obvious that $\overline{R(S,T)^{\circ}} = R(S,T)$. □

The next lemma follows trivially from our definitions.

LEMMA 4.4.5. *Let $R \in \mathcal{R}^1$.*

(1) $\cup_{S,T \in \mathcal{R}_2} R(S,T) = R$.
(2) *If $x \in R(S,T)^\circ_{m,n}$. If $n > 1$, there exists a unique $R' \in \mathcal{R}^1$ such that $f(x) \in R'(S,T)^\circ_{m+1,n-1}$. If $m > 1$, there exists a unique $R'' \in \mathcal{R}^1$ such that $f^{-1}(x) \in R''(S,T)^\circ_{m-1,n+1}$.*

LEMMA 4.4.6. *Let $R, R', R'' \in \mathcal{R}^1$, $S, T \in \mathcal{R}^2$.*

(s) *If $x \in R(S,T)$ and $f(x) \in R'(S,T)^\circ$, then $x \in R(S,T)^\circ$ and*
$$f(W^s(x, R(S,T)) \subset W^s(f(x), R'(S,T)).$$
(s) *If $x \in R(S,T)$ and $f^{-1}(x) \in R''(S,T)^\circ$, then $x \in R(S,T)^\circ$ and*
$$f^{-1}(W^u(x, R(S,T))) \subset W^u(f^{-1}(x), R''(S,T)).$$

PROOF. Let $f(x) \in R'(S,T)^{\circ\circ}$. It follows from Lemmas 4.4.3, 4.4.5 that $x \in R(S,T)^{\circ\circ}$ and $f(W^s(x, R(S,T)) \subset W^s(f(x), R'(S,T))$. If $f(x) \in R'(S,T)^\circ$, then $f(x) \in (R')^\circ$ and so $x \in R^\circ$. For some choice of $p, q > 0$, choose a sequence $x_n \to x$ such that $x_n \in R(S,T)^\circ_{p,q}$, $f(x_n) \in R'(S,T)^\circ_{p+1,q-1}$. By the previous argument, we have $f(W^s(x_n, R(S,T)))$ $\subset W^s(f(x_n), R'(S,T))$. Now let $n \to \infty$, proving (s). The proof of (u) is similar. □

REMARK 4.4.7. If, in the proof of Lemma 4.4.6, $x \in R^\circ$, $f(x) \in S$, it is not generally true that $f(W^s(x,R)) \subset W^s(f(x), S)$. However, this implication does follow if there exists a sequence $x_n \to x$ such that $(f(x_n)) \subset S^\circ$ (see, for example, [**38**, proof of Lemma 18.7.6]). ◇

LEMMA 4.4.8. *Let $R \in \mathcal{R}^1$, $S, T \in \mathcal{R}^2$, and $\gamma \neq \Gamma$.*

(1) $\gamma R(S,T) = (\gamma R)(\gamma S, \gamma T)$.
(2) $R(S,T) \cap \gamma R(S,T) \subset \Lambda_S$.

PROOF. It follows immediately from the definition of $R(S,T)^\circ_{m,n}$ that $\gamma R(S,T)^\circ_{m,n} = \{\gamma x \in \gamma R \mid f^{-m}(x) \in S^\circ, f^n(x) \in T^\circ\}$. Setting $\gamma x = y$ and using the Γ-invariance of the functions $M(x), N(x), f$ and the partition $\mathcal{R}$, it follows easily that $\gamma R(S,T)^\circ_{m,n} = (\gamma R)(\gamma S, \gamma T)^\circ_{m,n}$. Hence $\gamma R(S,T)^{\circ\circ} = (\gamma R)(\gamma S, \gamma T)^{\circ\circ}$ and, taking closures, (1) follows. Let $\gamma \in \Gamma$ and suppose $x \in R(S,T) \cap \gamma R(S,T) \setminus \Lambda_S$. Then there exists $m \in \mathbb{Z}$, such that $f^m(x) \in S \cup T$, and $|m|$ is minimal for this property. Without loss of generality, suppose $m > 0$ and so $f^m(x) \in T$. If $x \in \gamma R(S,T)$, then $f^m(x) \in \gamma T$. Since $T \in \mathcal{R}^2$, $\gamma T \cap T \neq \emptyset$ if and only $\gamma = e$, proving (2). □

LEMMA 4.4.9. *Let $S, T \in \mathcal{R}^2$. Suppose that $x \in S^\circ$, $f(x) \in R(S,T)^\circ$. Then*

(s) $f(W^s(x, S)) \subset W^s(f(x), R(S,T))$.

(u) $f^{-1}(W^u(f(x), R(S,T))) \subset W^u(x, S)$.

The corresponding inclusions hold if $x \in R(S,T)^\circ$, $f(x) \in T^\circ$.

PROOF. Suppose first that $f(x) \in R(S,T)^{\circ\circ}$. There exists $p > 0$ such that $f^p(x) \in T^\circ$. Since $\mathcal{R}$ is a Markov partition, $f^p(W^s(x, R) \subset W^s(f^p(x), T)$ and so we have $f(W^s(x, S)) \subset W^s(f(x), R(S,T))$. If we only have $f(x) \in R(S,T)^\circ$, we follow the argument of Lemma 4.4.6. The proof of (u) is immediate since $f^{-1}(W^u(f(x), R)) \subset W^u(x, S)$. □

Let $\mathcal{R}^{12} = \{R(S,T) \mid R \in \mathcal{R}^1, S, T \in \mathcal{R}^2, R(S,T) \neq \emptyset\}$. It follows from Lemmas 4.4.4, 4.4.5, 4.4.6, 4.4.9 that $\mathcal{M}' = \mathcal{R}^{12} \cup \mathcal{R}^2$ is a Markov partition on Λ. However, $\mathcal{M}'$ will not generally satisfy condition (b) for Γ-regular partitions. However, it follows from Lemma 4.4.8 that condition (b) is satisfied for all rectangles in $\mathcal{R}^{12}$.

To complete the proof of Theorem 4.4.1 it suffices to refine the partition $\mathcal{R}^2$. To this end, we define for each $R \in \mathcal{R}^1$, $S, T \in \mathcal{R}^2$

$$\begin{aligned} S(R;T) &= \overline{\{x \in S^\circ \mid f(x) \in R(S,T)^\circ\}}, \ T \in \mathcal{R}^2, R \in \mathcal{R}^1, \\ S_\Delta &= \overline{S \setminus \cup_{T \in \mathcal{R}^2, R \in \mathcal{R}^1} S(R;T)}. \end{aligned}$$

It is a routine exercise to verify that the sets $S(R;T)$, S_Δ are rectangles. Obviously $S_\Delta \cup_{R,T} S(R;T) = S$, for all $S \in \mathcal{R}^2$. Let $\mathcal{R}^{21} = \{S(R;T), S_\Delta \mid S, T \in \mathcal{R}^2, R \in \mathcal{R}^1\}$. It is straightforward to verify that the Markov intersection conditions hold between rectangles in $\mathcal{R}^{21}$ and rectangles in $\mathcal{R}^{12} \cup \mathcal{R}^{21}$. Hence $\mathcal{R}^{12} \cup \mathcal{R}^{21} = \mathcal{M}$ is a Markov partition. It follows from our constructions that $\mathcal{M}$ is Γ-regular.

CHAPTER 5

Transversally hyperbolic sets

Henceforth, we assume Γ is a non-finite compact Lie group acting smoothly on M. We follow the notations of 3.1.3.

5.1. Transverse hyperbolicity

DEFINITION 5.1.1. Let M be a riemannian Γ-manifold and $f \in \mathrm{Diff}_\Gamma(M)$. A compact f- and Γ-invariant subset Λ of M is *transversally hyperbolic for f* if the following conditions hold.

(a) All Γ-orbits in Λ have dimension equal to the dimension of Γ.

(b) There exists a Tf-invariant splitting $\mathbf{E}^s \oplus \mathbf{E}^u \oplus \mathbf{T}_\Lambda$ of $T_\Lambda M$ into continuous subbundles, and constants $c, C > 0$, $\lambda > 1 > \mu$, such that for all $n \in \mathbb{N}$,

$$\|T_x f^n(v)\| \leq c\mu^n \|v\|, \; (v \in \mathbf{E}^s_x, x \in \Lambda) \tag{5.1}$$

$$\|T_x f^n(v)\| \geq C\lambda^n \|v\|, \; (v \in \mathbf{E}^u_x, x \in \Lambda) \tag{5.2}$$

REMARKS 5.1.2. (1) A transversally hyperbolic set is *partially hyperbolic* in the sense of Brin & Pesin [**13**] (see also Pugh & Shub [**65**] and Lemma 5.1.3 below). We prefer the use of the term 'transversally hyperbolic' because the Γ-action automatically implies a highly regular center foliation (by Γ-orbits). The existence of a center foliation is not assumed in the definition of partial hyperbolicity. Indeed, a partially hyperbolic set may have no center foliation. (2) Condition (a) of Definition 5.1.1 is equivalent to requiring that all points in Λ have finite *isotropy group*. In fact, all the results of this section continue to hold if we weaken (a) to require only that all Γ-orbits of points in Λ have the *same* dimension (see Remark 3.1.5). ♢

LEMMA 5.1.3. *Suppose that Λ is transversally hyperbolic for f and let $\mathbf{E}^s \oplus \mathbf{E}^u \oplus \mathbf{T}_\Lambda$ denote the corresponding splitting of $T_\Lambda M$.*

(a) *We may choose a smooth Γ-equivariant riemannian metric on M with respect to which $Tf : \mathbf{T}_\Lambda \to \mathbf{T}_\Lambda$ is an isometry: $\|Tf(v)\| = \|v\|$, $(v \in \mathbf{T}_\Lambda)$.*

(b) *$\mathbf{E}^s$, $\mathbf{E}^u$ are unique, Hölder continuous and Γ-invariant subbundles of $T_\Lambda M$.*

PROOF. Granted (a), the asymptotic estimates (5.1,5.2) easily yield the uniqueness of the splitting. Similarly, the Γ-invariance of the splitting follows from (5.1,5.2) and the Γ-equivariance of f. For the Hölder continuity of the bundles see, for example, [**38**, Chapter 19, §1]. It remains to prove (a). Fix a riemannian metric on Γ which is invariant by left and right translations. This metric induces a metric on Γ/H for all closed subgroups H of Γ. In particular, if $\dim(\Gamma x) = \dim(\Gamma) = g$ for all $x \in \Lambda$, we may define a smooth family $\{d_\alpha | \alpha \in M(g)/\Gamma\}$ of riemannian metrics on the Γ-orbits in $M(g)$. Each metric d_α depends only on our original choice of metric on Γ. A straightforward partition of unity argument shows that we can construct a smooth Γ-invariant riemannian metric d on M such that for every Γ-orbit α in Λ, $d|\alpha$ extends d_α (view d, d_α as symmetric tensor fields on M, α respectively). In particular, α will be a totally geodesic submanifold of M for all $\alpha \in \Lambda/\Gamma$. With this choice of metric, $d(f(\gamma x), f(\eta x)) = d(\gamma x, \eta x)$ for all $x \in \Lambda$, $\eta, \gamma \in \Gamma$. Hence, Tf induces an isometry of $\mathbf{T}_\Lambda$. $\square$

Later we shall need results that imply a partially hyperbolic set is transversally hyperbolic. Before stating our main result, we need some preliminary definitions.

Suppose that $f \in \mathrm{Diff}_\Gamma(M)$ and Λ is a compact f- and Γ-invariant subset of the riemannian Γ-manifold M. Let $\mathbf{E}^s \oplus \mathbf{E}^u \oplus \mathbf{C}$ be a Tf-invariant splitting of $T_\Lambda M$ into continuous subbundles.

For $n \geq 1$, define

$$p_n = \inf_{x\in\Lambda} \inf_{\|v\|=1} \|T_x^n f(v)\|, \quad P_n = \sup_{x\in\Lambda} \sup_{\|v\|=1} \|T_x^n f(v)\|.$$

The constants p_n, P_n measure the strongest contraction and expansion of $Tf^n|\mathbf{C}$. Define

$$p = \liminf p_n^{\frac{1}{n}}, \quad P = \limsup P_n^{\frac{1}{n}}. \tag{5.3}$$

It follows from the definition of p, P that there exist constants $a, A > 0$ such that

$$ap^n\|v\| \leq \|T^n f(v)\| \leq AP^n\|v\|, (v \in \mathbf{C}, n \in \mathbb{N}).$$

DEFINITION 5.1.4. A splitting $\mathbf{E}^s \oplus \mathbf{E}^u \oplus \mathbf{C}$ of T_Λ is *partially hyperbolic* if we can find $\lambda > 1 > \mu$ satisfying (5.1,5.2) and

$$\lambda > P > p > \mu, \tag{5.4}$$

where P, p are defined by (5.3).

REMARKS 5.1.5. (1) It follows from (5.1,5.2,5.4) that if $\mathbf{E}^s \oplus \mathbf{E}^u \oplus \mathbf{C}$ is a partially hyperbolic splitting, then $\mathbf{E}^s$, $\mathbf{E}^u$ and $\mathbf{C}$ are Γ-invariant subbundles of $T_\Lambda M$. (2) It follows from Lemma 5.1.3 that if Λ is transversally hyperbolic, $P = p = 1$ and so the corresponding splitting is automatically partially hyperbolic. (3) In Definition 5.1.4, we have followed the definition of Pugh & Shub [**65**]. For the more general (and widely adopted) definition of Brin & Pesin, we refer to [**13**]. ◇

PROPOSITION 5.1.6. *Suppose that M be a riemannian Γ-manifold and $f \in \mathrm{Diff}_\Gamma(M)$. Let Λ be a compact f- and Γ-invariant subset of M and $\mathbf{E}^s \oplus \mathbf{E}^u \oplus \mathbf{C}$ be a partially hyperbolic splitting for $T_\Lambda M$. Let $d = \min_{x\in\Lambda} dim(\Gamma x)$. If d is greater than or equal to the fiber dimension D of $\mathbf{C}$, then $\mathbf{C} = \mathbf{T}_\Lambda$. In particular, if $dim(\Gamma) = D$, then Λ is transversally hyperbolic for f.*

PROOF. It is enough to prove that $\mathbf{C} = \mathbf{T}_\Lambda$. Let $\Lambda_d = M(d) \cap \Lambda$. Then Λ_d is a non-empty compact Γ- and f-invariant subset of Λ. Just as in the proof of Lemma 5.1.3, we may choose a smooth Γ-invariant riemannian metric on M relative to which Tf restricts to an isometry of $\mathbf{T}_{\Lambda_d}$. It follows from (5.1,5.2) that $\mathbf{C}|\Lambda_d = \mathbf{T}_{\Lambda_d}$. Suppose $\Lambda_d \neq \Lambda$. Let $\alpha = \Gamma x \subset \Lambda \setminus \Lambda_d$ be a Γ-orbit of dimension $d' > d$. Suppose there exist $\delta' > 0$ and a sequence $(n_i) \subset \mathbb{N}$ such that $d(f^{n_i}(\alpha), \Lambda_d) \geq \delta'$. Just as in the proof of Lemma 5.1.3, we can fix a metric on Γ and choose a smooth Γ-equivariant riemannian metric on M such that all Γ-orbits in $M(d')$ distance at least δ' from Λ_d have metric induced from that on Γ. Consequently, $f^{n_i} : \alpha \to f^{n_i}(\alpha)$ will be an isometry. Choosing a vector $v \in T\alpha \setminus \mathbf{C}|\alpha$, we obtain a contradiction using (5.1,5.2). A similar argument applies if there exist $\delta' > 0$ and a sequence $(n_i) \subset \mathbb{N}$ such that $\lim_{i\to-\infty} d(f^{n_i}(\alpha), \Lambda_d) \geq \delta'$. If neither of these conditions hold, we can choose a sequence $(n_i) \subset \mathbb{Z}$ such that $\lim_{i\to\pm\infty} d(f^{n_i}(\alpha), \Lambda_d) = 0$. Set $F = (\mathbf{E}^s_x \oplus \mathbf{E}^u_x) \cap \mathbf{T}_x$. Since $d' > d$, $\dim(F) \geq 1$. Set $F_i = T^{n_i}_x f(F)$, $x_i = f^{n_i}(x)$, $i \in \mathbb{Z}$. Either $F \subset E^s_x$ or not. If not, then as $i \to +\infty$, the angle between $E^u_{x_i}$ and F_i goes to zero. But this implies that f^{n_i} is stretching directions along group orbits as $i \to \infty$ and $f^{n_i}(\alpha)$ approaches Λ_d which is absurd. If $F \subset E^s_x$ then the same argument applies with f replaced by f^{-1}. Hence $\Lambda_d = \Lambda$. □

5.1.1. Examples of transversally hyperbolic sets. The simplest examples of transversally hyperbolic sets are provided by invariant group orbits of diffeomorphisms.

EXAMPLE 5.1.7. Suppose that α is an invariant Γ-orbit for f. Then α is normally hyperbolic if and only if α is transversally hyperbolic for f. A similar result holds if α is invariant by some power of f. We refer to [**20**, §3] for details. ♡

Most of our examples of transversally hyperbolic sets are based on the twisted product construction described in Examples 2.1.2(4).

Suppose that H is a finite subgroup of Γ and P is a smooth H-manifold. Let $f \in \mathrm{Diff}_H(P)$ and suppose that X is a compact H-invariant hyperbolic subset of P. Denote the associated Tf-invariant hyperbolic splitting of $T_X P$ by $\mathbf{E}^s \oplus \mathbf{E}^u$.

Define $M = P \times_H \Gamma$ and $\Lambda = X \times_H \Gamma$. Let $F = f_{\mathbf{e}} \in \mathrm{Diff}_\Gamma(M)$, where $\mathbf{e}$ is the 'identity' cocycle.

LEMMA 5.1.8. *The set $\Lambda \subset M$ is transversally hyperbolic for F.*

PROOF. Regard X and P as equivariantly embedded in M by the map $\iota(x) = [x, e]$, $x \in P$. Since X is an H-invariant hyperbolic subset of P, we have a splitting $T_X P = \mathbf{E}^s \oplus \mathbf{E}^u$, where $\mathbf{E}^s$, $\mathbf{E}^u$ are continuous H-invariant subbundles of $T_X P$. The bundles $\mathbf{E}^s, \mathbf{E}^u$ extend Γ-equivariantly to Λ and $\mathbf{E}^s \oplus \mathbf{E}^u \oplus \mathbf{T}_\Lambda = T_\Lambda M$ is the required transversally hyperbolic splitting for F. □

5.2. Properties of transversally hyperbolic sets

Assume that a fixed choice of smooth Γ-invariant riemannian metric has been made for M. Let $d(\ ,\ )$ denote the associated Γ-invariant metric on M.

The results we describe often follow from general results in [**37**, §7]. However, we sometimes give direct (simple) proofs that use the presence of the group action and the smooth foliation by group orbits.

The stable manifold theory for transversally hyperbolic sets follows from straightforward equivariant versions of the results in [**37**] (see also [**20, 13**]). Specifically, suppose that $\alpha \subset \Lambda$ is a Γ-orbit. There exist equivariantly smoothly immersed stable and unstable manifolds $W^s(\alpha), W^u(\alpha)$ for Λ through α. Along α, the stable manifold is tangent to $\mathbf{E}^s \oplus \mathbf{T}$, the unstable manifold to $\mathbf{E}^u \oplus \mathbf{T}$. In particular, $W^s(\alpha)$ intersects $W^u(\alpha)$ transversally along α. Let $\varepsilon > 0$ and define local stable and unstable manifolds through α by

$$\begin{aligned} W^s_\varepsilon(\alpha) &= \{y \in W^s(\alpha) \mid d(f^n(\alpha), f^n(y)) \le \varepsilon,\, n \in \mathbb{N}\}, \\ W^u_\varepsilon(\alpha) &= \{y \in W^u(\alpha) \mid d(f^{-n}(\alpha), f^{-n}(y)) \le \varepsilon,\, n \in \mathbb{N}\}. \end{aligned}$$

Both $W^s_\varepsilon(\alpha)$ and $W^u_\varepsilon(\alpha)$ will be embedded Γ-invariant submanifolds of M, depending continuously on $\alpha \subset \Lambda$. For sufficiently small $\varepsilon > 0$, the intersection of $W^s_\varepsilon(\alpha)$ and $W^u_\varepsilon(\alpha)$ will consist of a finite number of Γ-orbits, all Γ-orbits α (obviously, for a specific orbit α, we can choose $\varepsilon > 0$ such that $W^s_\varepsilon(\alpha) \cap W^u_\varepsilon(\alpha) = \alpha$).

Unless the action of Γ on Λ is free, it is not possible to remove the condition $y \in W^{s/u}(\alpha)$ from the definitions of the local stable and unstable manifolds of α.

EXAMPLE 5.2.1. Let $\Lambda \subset \mathbb{R}_2$ denote the $\mathbb{Z}_2(\kappa)$-invariant full shift on the symbols $\{-1, 0, 1\}$ as described in Example 4.2.12. If $(x_i), (y_i) \in \Lambda$, we define $d((x_i), (y_i))$ to be 2^{-j}, where j is the smallest positive integer such either that $x_j = y_j$, or $x_{-j} = y_{-j}$. Given $\varepsilon > 0$, choose $p \in \mathbb{N}$ such that $2^{-p} < \varepsilon$ and set $m = 2p + 1$. Consider $\mathbf{x} \in \Lambda$ of the form $\mathbf{x} = \ldots \mathbf{a}_n \mathbf{0}_{m_n} \mathbf{a}_{n+1} \ldots$, where

(a) $\mathbf{0}_{m_n}$ is a string of m_n zeros, where $m_n \geq m$, all $n \in \mathbb{Z}$.
(b) Each $\mathbf{a}_n$ is a non-empty finite string of nonzero symbols.

Set $\kappa \mathbf{a}_n = \bar{\mathbf{a}}_n$ and note that $\bar{\mathbf{a}}_n \neq \mathbf{a}_n$. Let $\mathbf{y} = \ldots \mathbf{b}_n \mathbf{0}_{m_n} \mathbf{b}_{n+1} \ldots$, where $\mathbf{b}_n \in \{\mathbf{a}_n, \bar{\mathbf{a}}_n\}$. It follows by our construction that $d(\sigma^j \mathbf{y}, \mathbb{Z}_2 \sigma^j \mathbf{x}) < \varepsilon$, $j \in \mathbb{Z}$. However, for 'most' choices of $\mathbf{y}$, $\mathbf{y} \notin W^s(\mathbb{Z}_2 \mathbf{x}) \cap W^u(\mathbb{Z}_2 \mathbf{x})$. ♡

PROPOSITION 5.2.2. *For sufficiently small $\varepsilon > 0$, independent of α,*

$$\begin{aligned} W^s_\varepsilon(\alpha) &= \{y \mid \exists x \in \alpha, d(f^n(x), f^n(y)) \leq \varepsilon,\ n \in \mathbb{N}\}, \\ W^u_\varepsilon(\alpha) &= \{y \mid \exists x \in \alpha, d(f^n(x), f^n(y)) \leq \varepsilon,\ -n \in \mathbb{N}\}. \end{aligned}$$

PROOF. The proof is similar to that of the corresponding result when there is no group action. □

The invariant manifolds of Γ-orbits are foliated by the strong stable and unstable manifolds in the usual way. Thus, if $x \in \alpha$, the strong stable manifold of x is $W^{ss}(x) = \{y \in W^s(\alpha) \mid d(f^n(x), f^n(y)) \to 0, n \to \infty\}$. We have

$$W^{ss}(\gamma x) = \gamma W^{ss}(x),\ \ (x \in \alpha, \gamma \in \Gamma),\ \ W^s(\alpha) = \cup_{x \in \alpha} W^{ss}(x).$$

We similarly define the strong unstable foliation $\{W^{uu}(x) \mid x \in \alpha\}$ of $W^u(\alpha)$. We have $T_x W^s(x) = \mathbf{E}^s_x$, $T_x W^u(x) = \mathbf{E}^u_x$. For all $x \in \Lambda$, $W^{ss}(x)$ intersects $W^u(\Gamma x)$ transversally at x. Similarly, $W^{uu}(x)$ intersects $W^s(\Gamma x)$ transversally at x. For $\varepsilon > 0$, and $x \in \alpha$, we define $W^{ss}_\varepsilon(x) = W^s_\varepsilon(\alpha) \cap W^{ss}(x)$, $W^{uu}_\varepsilon(x) = W^u_\varepsilon(\alpha) \cap W^{uu}(x)$.

REMARK 5.2.3. If $x \in \alpha$, $y \in M$ and $d(f^n(x), f^n(y)) \leq \varepsilon$, $n \in \mathbb{N}$, there exists $x' \in \alpha$, $d(x, x') \leq \varepsilon$, such that $y \in W^{ss}(x')$ (Proposition 5.2.2). ◊

The next result will be useful when we define local product structures for transversally hyperbolic sets.

PROPOSITION 5.2.4. *Let Λ be transversally hyperbolic. We may choose $a, \varepsilon > 0$ and an open neighborhood A of $e \in \Gamma$ such that if we define $U = \{(x, y) \in \Lambda^2 \mid d(x, y) < a\}$ then there exist Hölder continuous maps $[\![\,,\,], [\,,\,]\!] : U \to M$ and $\rho_u, \rho_s : U \to A$ such that for* *all* $(x,y) \in U$,

$$\begin{aligned} [\![x,y] &= W^{ss}_\varepsilon(x) \cap W^{uu}_\varepsilon(\rho_u(x,y)y), \\ &= W^{ss}_\varepsilon(x) \cap W^{u}_\varepsilon(Ay), \\ [x,y]\!] &= W^{ss}_\varepsilon(\rho_s(x,y)x) \cap W^{uu}_\varepsilon(y), \\ &= W^{s}_\varepsilon(Ax) \cap W^{uu}_\varepsilon(y). \end{aligned}$$

PROOF. For $s > 0$, let A denote the open d_Γ s-disk neighborhood of $e \in \Gamma$. For $x \in \Lambda$, $r > 0$, let N_x denote the orthogonal complement of $T_x\Gamma x$ in T_xM and $D_x(r)$ denote the open r-disk, center 0 in N_x. Set $D_x = \exp(D_x(r))$. Then we may choose $r > 0$, such that for all $x \in \Lambda$ and $y \in D_x$, $\Gamma y \pitchfork D_x$ and the intersection consists of finitely many points. Since the disks D_x are transverse to group orbits and all isotropy groups are finite, we may choose $s > 0$ sufficiently small so that $D_x \cap D_{\gamma x} = \emptyset$, for all $\gamma \in A \setminus \{e\}$, $x \in \Lambda$. Just as in the case when there is no group action, we may choose $a, \varepsilon > 0$ such that if $U = \{(x,y) \in \Lambda^2 \mid d(x,y) < a\}$, then for all $(x,y) \in U$, $W^{ss}_\varepsilon(x) \pitchfork W^u_\varepsilon(Ay)$, $W^s_\varepsilon(Ay) \pitchfork W^{uu}_\varepsilon(x)$ and both intersections consist of a single point contained in $A(S_x)$. It follows from standard results on transversality that $W^{ss}_\varepsilon(x) \cap W^u_\varepsilon(Ay) = W^{ss}_\varepsilon(x) \cap W^{uu}(\rho_u(x,y)y)$, where $\rho_u(x,y) \in A$ depends continuously on $(x,y) \in U$. Similarly for the other intersection. Finally, for the Hölder continuity statement, we refer to [**38**, Chapter 19]. □

REMARK 5.2.5. In general, if $(x,y) \in U$ then $W^{ss}_\varepsilon(x) \cap W^u_\varepsilon(\Gamma y)$ will consist of more than one point. The number of points in the intersection will be bounded by the maximal order of isotropy group for the action of Γ on Λ. Points in $W^{ss}_\varepsilon(x) \cap W^u_\varepsilon(\Gamma y)$ will lie on distinct Γ-orbits. ◇

In the sequel we adopt the following convention. Whenever we write $W^{s/u}_\varepsilon(\Gamma x)$ or $W^{ss/uu}_\delta(x)$, we always assume that $\varepsilon, \delta > 0$ are chosen sufficiently small so that the conditions of Proposition 5.2.4 are satisfied.

5.3. Γ-expansiveness

Let $\rho > 0$ and let A_ρ denote the closed ρ-disk, center e, in Γ. Set $\mathcal{P}_\rho = \{A_\rho(x) \mid x \in \Lambda\}$. We regard $\mathcal{P}_\rho$ as a *plaquation* of Λ (strictly, of the associated lamination of Λ [**37**, §7]).

We recall that a sequence $\mathbf{x} = (x_i)_{i=-\infty}^{\infty} \subset M$ is an *α-pseudo-orbit* if

$$d(x_{i+1}, f(x_i)) < \alpha, \quad (i \in \mathbb{Z}).$$

DEFINITION 5.3.1 (cf [**37**, §7]). We say that an α-pseudo-orbit $\mathbf{x}$ respects $\mathcal{P}_\rho$ if $f(x_i) \in A_\rho(x_{i+1})$, all i.

DEFINITION 5.3.2 (cf [**37**, §7]). The map $f : \Lambda \to \Lambda$ is Γ-*expansive* if there exist $\alpha, \rho > 0$ such that if $\mathbf{x}, \mathbf{y}$ are α-pseudo-orbits respecting $\mathcal{P}_\rho$, and $d(x_i, y_i) < \varepsilon$, all i, then $x_i \in A_\rho(y_i)$, all i. In particular, $\mathbf{x}$, $\mathbf{y}$ determine the same sequence in Λ/Γ.

THEOREM 5.3.3. *If Λ is transversally hyperbolic then $f : \Lambda \to \Lambda$ is Γ-expansive.*

PROOF. In the terminology of [**37**], $f : \Lambda \to \Lambda$ is 0-normally hyperbolic to the lamination of Λ by Γ-orbits. Since this lamination is trivially smoothable, the result follows from [**37**, Theorem 7.4]. Alternatively, it is easy to give a simple direct proof. □

COROLLARY 5.3.4. *Let Λ be transversally hyperbolic. There exist an open Γ-invariant neighborhood U of Λ in M and constants $\beta, \rho > 0$ such that whenever $x \in \Lambda$, $y \in U$ and there exists a sequence $(\tau_n) \subset \Gamma$ such that for all $n \in \mathbb{Z}$ we have*

(a) $d_\Gamma(\tau_n, \tau_{n+1}) < \rho$,
(b) $d(f^n(x), \tau_n f^n(y)) < \beta$,

then $y \in \Gamma x$.

PROOF. Our condition on y implies that the f-orbit of y stays close to Λ. Just as in [**38**, Proposition 6.4.6], we may choose an open Γ-invariant neighborhood U of Λ such that if $\Lambda' = \cap_{n \in \mathbb{Z}} f^n(\bar{U})$, then $T_{\Lambda'}$ has a partially hyperbolic splitting. It follows from Proposition 5.1.6 that Λ' is transversally hyperbolic. Hence it is no loss of generality to take $\Lambda = \Lambda'$ and assume that $y \in \Lambda$. Now apply Theorem 5.3.3 with $x_i = f^i(x)$, $y_i = \tau_i f^i(y)$, $i \in \mathbb{Z}$. □

5.4. Stability properties of transversally hyperbolic sets

In 5.8, we give a sufficient condition for the persistence of transversal hyperbolicity under perturbations within $\mathrm{Diff}_\Gamma(M)$. For now we just verify stability of transversal hyperbolicity under perturbations by elements of $\mathcal{D}_I(M)$.

Let $f \in \mathrm{Diff}_\Gamma(M)$ and Λ be a compact f- and Γ-invariant subset of M. We define the linear isomorphism $f_\star : C^0_\Gamma(T_\Lambda M) \to C^0_\Gamma(T_\Lambda M)$ by $f_\star(X) = Tf \circ X \circ f^{-1}$, $X \in C^0_\Gamma(T_\Lambda M)$. Observe that $f_\star(\mathcal{T}^0(\Lambda)) = \mathcal{T}^0(\Lambda)$. If all Γ-orbits in Λ are of dimension equal to that of Γ, then we can choose a Γ-invariant riemannian metric on M so that Tf restricts to an isometry on $\mathbf{T}_\Lambda$. It follows that the spectrum of $f_\star|\mathcal{T}^0(\Lambda)$ lies on the unit circle.

PROPOSITION 5.4.1. *Let f, Λ be as above and suppose that all Γ-orbits in Λ have dimension $g = dim(\Gamma)$. Then Λ is transversally hyperbolic if and only if there exists a splitting $\mathcal{T}^0(\Lambda) \oplus \mathcal{K}$ of $C^0_\Gamma(T_\Lambda M)$ with respect to which $f_\star$ has matrix $\begin{pmatrix} A & B \\ 0 & F \end{pmatrix}$, where A has spectrum on the unit circle and the spectrum of $F : \mathcal{K} \to \mathcal{K}$ is disjoint from the unit circle.*

PROOF. The proof is a straightforward equivariant generalization of the corresponding well-known characterization of hyperbolic sets (no group action). (See also [**23**, §2]). □

THEOREM 5.4.2. *Let $f \in \mathrm{Diff}_\Gamma(M)$ and suppose that $\Lambda \subset M$ is compact and f- and Γ-invariant. If there exists $g \in \mathcal{D}_f(M)$ such that Λ is transversally hyperbolic for g, then Λ is transversally hyperbolic for all $h \in \mathcal{D}_f(M)$.*

PROOF. Without loss of generality, suppose Λ is transversally hyperbolic for f. Let $g = \eta \circ f$, $\eta \in \mathcal{D}_I(M)$. A straightforward computation verifies that $g_\star$ satisfies the hypotheses of Proposition 5.4.1. □

EXAMPLE 5.4.3 (Skew products). Let M be a trivial Γ-manifold and consider the skew product $M \times \Gamma$. Given $f \in \mathrm{Diff}(M)$ and a smooth cocycle $\phi : M \to \Gamma$, define $f_\phi \in \mathrm{Diff}_\Gamma(M \times \Gamma)$ by

$$f_\phi(\gamma, x) = (\gamma\phi(x), f(x)), \quad (\gamma \in \Gamma, x \in M)$$

Thus f_ϕ is a Γ-*extension of* f or the *skew product of* f *with* Γ. It follows from Theorem 5.4.2 that if $X \subset M$ is hyperbolic then $\Gamma \times X$ is transversally hyperbolic for all skew extensions of f. Similar remarks hold for principal bundle extensions of f. ♡

5.5. Subshifts of finite type and attractors

It is straightforward to generalize the definitions of subshift of finite type to include compact Lie group actions [**22**]. We briefly indicate some of the main points.

We say that a Γ-equivariant homeomorphism $F : \Lambda \to \Lambda$ is a Γ-subshift of finite type if we can find a finite subgroup $H \subset \Gamma$, an H-equivariant subshift of finite type $\sigma : X \to X$ and a continuous skew H-equivariant map $\phi : X \to H$ such that $\Lambda \cong X \times_H \Gamma$ and, relative to this isomorphism, $F = \sigma_\phi$.

It may shown that every Γ-subshift of finite type can be realized as a finite union of indecomposable pieces of the Ω-set of a Γ-equivariantly structurally stable equivariant diffeomorphism [**22**].

We can describe two large classes of transversally hyperbolic attractors for Γ-equivariant diffeomorphisms. The first class consists of (twisted) products or skew extensions of the hyperbolic attractors described in 4.3.2. A second class may be based on twisted products with Anosov diffeomorphisms. Details are given in [**23**].

EXAMPLES 5.5.1. (1) Let $\Lambda \subset \mathbb{R}^3$ be the $\mathbb{Z}_2$-invariant solenoid for the $\mathbb{Z}_2$-equivariant diffeomorphism F of $\mathbb{R}^3$ that was constructed in Example 4.3.5. Embed $\mathbb{Z}_2$ as the center of SU(2) (generated by $-I$). Let $\phi : \mathbb{R}^3 \to \mathrm{SU}(2)$ be a smooth $\mathbb{Z}_2$-invariant map. Since $\mathbb{Z}_2$ is the center of SU(2), ϕ is $\mathbb{Z}_2$-equivariant and so F_ϕ defines an SU(2)-equivariant diffeomorphism of the twisted product $\mathbb{R}^3 \times_{\mathbb{Z}_2} \mathrm{SU}(2)$. The set $\Lambda \times_{\mathbb{Z}_2} \mathrm{SU}(2)$ is a transversally hyperbolic attractor for F_ϕ which has countably many singular SU(2)-orbits.
(2) Let $T : T^2 \to T^2$ be the Thom-Anosov map of the torus. Consider the $\mathbb{Z}_2$-action induced on T^2 by $-Id : \mathbb{R}^2 \to \mathbb{R}^2$. Observe that $\mathbb{Z}_2$ has four fixed points. Regard $\mathbb{Z}_2$ as the subgroup of $\mathbb{K}$ generated by $\exp(\imath\pi)$. Let $\phi : T^2 \to \mathbb{K}$ be a smooth cocycle. Then $T_\phi \in \mathrm{Diff}_{\mathbb{K}}(T^2 \times_{\mathbb{Z}_2} \mathbb{K})$. Since T is Anosov (hyperbolic), T_ϕ is transversally hyperbolic. We note that $T^2 \times_{\mathbb{Z}_2} \mathbb{K}$ has 4 singular $\mathbb{K}$-orbits. ♡

5.6. Local product structure

Let M be a riemannian Γ-manifold with associated metric d. Let $\exp : TM \to M$ denote the exponential map of M. We assume (see Remark 3.1.5) that (a) Γ acts freely on the principal orbit stratum M_N of M, and (b) all Γ-orbits are of the same dimension (equivalently, all isotropy groups are finite).

Let $f \in \mathrm{Diff}_\Gamma(M)$ and suppose that $\Lambda \subset M$ is transversally hyperbolic and $\Lambda \cap M_N = \Lambda_N$ is dense in Λ.

DEFINITION 5.6.1. The transversally hyperbolic set Λ has *local product structure* if we can choose a Γ-invariant open neighborhood U_Δ of $\Delta(\Lambda) \subset \Lambda^2$ such that $[\![x, y]\!] \in \Lambda$, all $(x, y) \in U_\Delta$. (Equivalently, if $[\![\ ,\ ]\!]$ takes values in Λ.)

REMARKS 5.6.2. (1) An equivalent formulation of local product structure is that $W^s_\varepsilon(\Gamma x) \cap W^u_\varepsilon(\Gamma y) \subset \Lambda$, for all $(x, y) \in U_\Delta$. (2) Our definition of local product structure implies ε-local product structure [**37**, §7A]. ♢

DEFINITION 5.6.3. Let $f \in \mathrm{Diff}_\Gamma(M)$. A Γ-invariant subset Λ of M is a *transversally hyperbolic basic set*, or just *basic*, for f if

(a) Λ is a compact f-invariant set.

(b) Λ is a transversally hyperbolic set for f.
(c) Λ has local product structure.
(d) $\tilde{f} : \Lambda/\Gamma \to \Lambda/\Gamma$ has a dense orbit (topological transitivity).

5.7. Expansiveness and shadowing

DEFINITION 5.7.1. Let $\beta > 0$ and $\mathbf{x} = (x_i)_{i=-\infty}^{\infty} \subset M$. Let $\rho \in (0, \rho_0]$. We say that the f-orbit of $x \in M$ *equivariantly (β, ρ)-shadows* $\mathbf{x}$ if there exists a sequence $(\gamma_i) \subset A_\rho$ such that

$$d(f^i(x), \Pi_{j=0}^{i} \gamma_j x_i) < \beta, \quad (i \in \mathbb{Z}).$$

REMARK 5.7.2. If Γ is finite, then equivariant (β, ρ)-shadowing is just β-shadowing. ◇

PROPOSITION 5.7.3. *Let Λ be a basic set of $f \in \mathrm{Diff}_\Gamma(M)$. For every $\beta, \rho > 0$, there exists $\alpha > 0$, and a Γ-invariant open neighborhood U of Λ in M such that every α-pseudo-orbit $\mathbf{x} \subset U$ is equivariantly (β, ρ)-shadowed by the f-orbit of a point $\theta(\mathbf{x}) \in \Lambda$.*

PROOF. Using Proposition 5.2.4, Bowen's (original) proof of the shadowing theorem [**9**, 3.6] extends without difficulty to prove the existence of (β, ρ)-shadowing orbits. Alternatively, one may apply [**37**, 7A.2]. □

Of course, when there is no group action it follows from expansiveness that for sufficiently small $\beta > 0$, the shadowing orbit $\theta(\mathbf{x})$ is unique. We shall show for sufficiently small $\beta, \rho > 0$, $\theta(\mathbf{x})$ is unique (mod Γ).

THEOREM 5.7.4. *Let Λ be a basic set of $f \in \mathrm{Diff}_\Gamma(M)$. For sufficiently small $\beta, \rho > 0$, there exists $\alpha > 0$, and a Γ-invariant open neighborhood U of Λ in M such that every α-pseudo-orbit $\mathbf{x} \subset U$ is equivariantly (β, ρ)-shadowed by the f-orbit of a unique (mod Γ) point $\theta(\mathbf{x})$. Furthermore, $\theta(\mathbf{x}) \in \Lambda$.*

PROOF. If the f-orbits of y_1, y_2 equivariantly (β, ρ)-shadow an α-pseudo-orbit $\mathbf{x}$, then clearly the f-orbit of y_1 equivariantly $(2\beta, 2\rho)$-shadows the orbit $(f^n(y_2))$. Using Proposition 5.7.3, it is no loss of generality to assume $y_1 \in \Lambda$. For sufficiently small (β, ρ) we may apply Corollary 5.3.4 to deduce that $y_2 \in \Gamma y_1$. □

THEOREM 5.7.5. *Let Λ be a basic set for $f \in \mathrm{Diff}_\Gamma(M)$. Then periodic points of $\tilde{f} : \Lambda/\Gamma \to \Lambda/\Gamma$ are dense in Λ/Γ.*

PROOF. Let $x \in \Lambda$ and $\delta > 0$. It suffices to find $y \in \Lambda$ such that Γy is periodic (for $\tilde{f}$) and $d(\Gamma x, \Gamma y) < \delta$. Fix $\alpha, \beta, \rho > 0$ so that we have uniqueness of (β, ρ)-orbits shadowing α-pseudo-orbits in Theorem 5.7.4. We may and shall assume that $\alpha + \beta < \delta$.

Since $\tilde{f}$ is topologically transitive we can pick a point $z \in \Lambda$ (of trivial isotropy) such that the $\tilde{f}$-orbit of Γz is dense in Λ/Γ and $d(x, z) < \alpha$. Choose $n > 0$ so that $d(\Gamma z, \Gamma f^n(z)) < \alpha$. Composing f with an element of $\mathcal{D}_I(M)$, we may assume that $d(z, f^n(z)) < \alpha$ (if Γ is not connected, we may have to replace f^n by f^{np}, where p is bounded by the number of connected components of Γ). Let $\mathbf{z}$ be the α-pseudo-orbit defined by

$$\begin{aligned} z_i &= f^i(z), \quad 0 \leq i < n, \\ &= z_j, \ j = i \bmod n, \quad i < 0, i \geq n. \end{aligned}$$

Let $y = \theta(\mathbf{z})$ and $(\gamma_i) \subset A_\rho$ be the associated sequence given by Theorem 5.7.4. For $n \in \mathbb{Z}$, set $\tau_n = \Pi_{j=0}^n \gamma_j$. For $i \in \mathbb{Z}$, we have

$$d(\tau_i f^i(y), z_i) \leq \beta, \quad d(\tau_{i+n} f^{i+n}(y), z_{i+n}) \leq \beta.$$

Since $z_{i+n} = z_i$, it follows from the second inequality that

$$d(\tau_i' f^i(f^n(\xi y)), z_i) \leq \beta,$$

where $\xi = \tau_n^{-1}$ and $\tau_i' = \tau_{i+n}\tau_n^{-1}$, $i \in \mathbb{Z}$. It follows that the f-orbit of $f^n(\xi y)$ (β, ρ)-shadows $\mathbf{z}$. Hence, by uniqueness of shadowing, $\Gamma y = \Gamma f^n(y)$. Clearly, $d(\Gamma x, \Gamma y) \leq \alpha + \beta < \delta$. □

DEFINITION 5.7.6. A Γ- and f-invariant subset Λ of $f \in \mathrm{Diff}_\Gamma(M)$ *locally maximal* if there exists a closed neighborhood V of Λ such that $\cap_{n \in \mathbb{Z}} f^n(V) = \Lambda$.

THEOREM 5.7.7. *Suppose that Λ is a Γ-invariant, compact transversally hyperbolic set for $f \in \mathrm{Diff}_\Gamma(M)$. Then Λ is locally maximal if and only if Λ has local product structure.*

PROOF. It is trivial (just as in the case where there is no group action) that locally maximality implies that Λ has local product structure. The proof of the reverse implication follows that in [**50**, Corollary 3.6]. Choose $1 > \beta, \rho > 0$ small and $\alpha \in (0, \beta)$ such that every α-pseudo-orbit in Λ is uniquely (mod Γ) (β, ρ)-shadowed by an orbit in Λ. Choose $\delta \in (0, \alpha/2)$ so that $d(x, y) < \delta$ implies $d(f(x), f(y)) < \alpha/2$. Let $V = \{y \in M \mid d(y, \Lambda) \leq \delta\}$. If $x \in \cap_{n \in \mathbb{Z}} f^n(V)$, then for each $n \in \mathbb{Z}$, there exists $x_n \in \Lambda$ such that $d(f^n(x), x_n) \leq \delta$. For $n \in \mathbb{Z}$, we have

$$\begin{aligned} d(f(x_n), x_{n+1}) &\leq d(f(x_n), f^{n+1}(x)) + d(f^{n+1}(x), x_{n+1}), \\ &\leq \alpha/2 + \delta \leq \alpha. \end{aligned}$$

Hence (x_n) is an α-pseudo-orbit. By Theorem 5.7.4, there exists an orbit $(f^n(y)) \subset \Lambda$ which (β, ρ)-shadows (x_n). It follows by uniqueness of shadowing that $x \in \Gamma y$. □

5.8. Stability of basic sets

We conclude this chapter with a stability theorem for basic sets.

THEOREM 5.8.1. *Let Λ be a basic set for $f \in \mathrm{Diff}_\Gamma(M)$. There are neighborhoods $\mathcal{U}$ of $f \in \mathrm{Diff}_\Gamma(M)$ (C^1-topology), and U of Λ in M such that if $g \in \mathcal{U}$, then $\Lambda(g) = \cap_{n\in\mathbb{Z}} g^n U$ is a basic set for g. Further, there is a homeomorphism $\phi_g : \Lambda/\Gamma \to \Lambda(g)/\Gamma$ such that $g\phi_g = \phi_g f$ and ϕ_g depends continuously on $g \in \mathcal{U}$.*

PROOF. Our proof of Theorem 5.8.1 is modeled on the presentation of the corresponding result for non-equivariant systems given by Newhouse [**50**, Theorem 3.7] (an alternative approach may be based on [**37**]). Just as in [**50**], we may choose a closed neighborhood U of Λ such that $\cap_{n\le 0} f^n(U) = W^s_e ps(\Lambda)$, $\cap_{n\ge 0} f^n(U) = W^u_e ps(\Lambda)$ and $\cap_{n\in\mathbb{Z}} f^n(U) = \Lambda$. If g is C^1 close to f, then $\Lambda(g) = \cap_{n\in\mathbb{Z}} g^n(U)$ is a transversally hyperbolic set for g. Since $\Lambda(g) \subset U^\circ$. it follows that $\Lambda(g)$ has a local product structure. In order to construct the conjugacy ϕ_g, we use Theorem 5.7.4 just as the shadowing lemma is used in the proof of [**50**, Theorem 3.7]. The only difference is that the conjugacy is now only well-defined on Γ-orbits. □

REMARK 5.8.2. General stability results for partially hyperbolic sets are given in [**37**, §7]. See also [**61**, Theorem 5] for attractors. ◊

CHAPTER 6

Markov partitions for basic sets

6.1. Rectangles

Throughout this section we suppose that Λ is a basic set for $f \in \mathrm{Diff}_\Gamma(M)$. Let $U_\Delta = \{(x,y) \in \Lambda \mid d(x,y) < a\}$ denote the neighborhood of the diagonal in Λ^2 on which $[\![\ ,\ ],[\ ,\ ]\!]$ are defined and continuous (Proposition 5.2.4).

Suppose that $X \subset M$ is a compact Γ-invariant set. We define $\delta(X) = \sup\{d(\Gamma x, \Gamma y) \mid x, y \in X\}$ – the diameter of X transverse to the Γ-action.

DEFINITION 6.1.1. A closed Γ-invariant subset R of Λ is a *rectangle* if

(R0) $\delta(R) < a$.
(R1) $R = \overline{R^\circ}$.
(R2) There exists $x_0 \in R$ such that $(\Gamma_y) \leq (\Gamma_{x_0})$, all $y \in R$.
(R3) For all $x \in R^\circ$, $y \in W^{uu}_\varepsilon(x) \cap R$, $z \in W^{ss}_\varepsilon(x) \cap R$, $[\![y, z] \in R$.

If, in addition, we have

(RS3) $x, y \in R^2 \cap U_\Delta \Longrightarrow [\![x, y] \in R$,

(equivalently, $[x, y]\!] \in R$), we say that R is a *strict rectangle*.

Following (R2), we let $\rho(R) \in \mathcal{O}(\Lambda)$ denote the unique maximal isotropy type of points in R. In the sequel, we refer to $\rho(R)$ as the symmetry type of R. Set $|\rho(R)| = |G|$, where $G \in \rho(R)$. If $\rho(R)$ is trivial ($|\rho(R)| = 1$), all points in R have the same trivial isotropy type.

EXAMPLE 6.1.2. Let $x \in \Lambda$ and $A \subset W^{uu}_\varepsilon(\Gamma x)$, $B \subset W^{ss}_\varepsilon(\Gamma x)$ be closed Γ_x-invariant r-disk neighborhoods of x. Let $R = \Gamma[\![A, B]$. Provided $r > 0$ is sufficiently small, R is a strict rectangle of symmetry type (Γ_x). ♡

LEMMA 6.1.3. *Let R be a strict rectangle. Choose $\delta > 0$ satisfying $0 < \delta(R) \ll \delta < a$. Then for all $x, y \in R$,*

$$W^s_\delta(\Gamma x) \cap W^u_\delta(\Gamma y) \subset R.$$

PROOF. Suppose $z \in W^s_\delta(\Gamma x) \cap W^u_\delta(\Gamma y)$. Then $z \in W^{ss}_\delta(\gamma x) \cap W^u_\delta(\eta y)$ for some $\gamma, \eta \in \Gamma$. That is, $z = [\![\gamma x, \eta y] \in R$. □

Given $\delta > 0$, and a rectangle R, define

$$W^s(\Gamma x, R) = W^s_\delta(\Gamma x) \cap R, \quad W^u(\Gamma x, R) = W^u_\delta(\Gamma x) \cap R.$$

(It is assumed that $0 < \delta(R) \ll \delta$.)

6.2. Slices

Let $\exp : TM \to M$ denote the exponential map of the given Γ-invariant riemannian metric on M. It follows by the equivariant tubular neighborhood theorem that for each $x \in \Lambda$, we can find a greatest $r = r(x) > 0$ such that exp Γ-equivariantly embeds the open r-disk subbundle $N(r)$ of the normal bundle $N = T\Gamma x^\perp$ of Γx as a tubular neighborhood of Γx. For every $y \in \Gamma x$, $\exp(N(r)_y) = D(y, r)$ will be a slice at y of diameter $2r$.

Recall that we have a partition $\{\Lambda_i \mid 1 \le i \le N\}$ of Λ into points of the same submaximal isotropy type. In particular, Λ_1 consists of points of maximal isotropy type for the action of Γ on Λ and $\Lambda_N = \Lambda_\Pi$ is the open dense subset of Λ on which Γ acts freely. We define closed Γ-invariant subsets Λ^k of Λ by

$$\begin{aligned} \Lambda^k &= \emptyset, \ k = 0, \\ &= \cup_{j=1}^k \Lambda_j, \ 1 \le k < N. \end{aligned}$$

For $x \in \Lambda_k$, define $d_k(x) \in \mathbb{R}$ by

$$\begin{aligned} d_k(x) &= 1, \ k = 1, \\ &= d(x, \Lambda^{k-1}), \ 2 \le k \le N. \end{aligned}$$

LEMMA 6.2.1. *There exist constants* $c_1, \ldots, c_N > 0$ *such that if* $x \in \Lambda_k \setminus \Lambda^{k-1}$, *then* $D(x, c_k d_k(x))$ *is a slice at* x.

PROOF. Since all group orbits are of the same dimension it is easy to choose $r > 0$ such that $\exp|N(r)_x$ is transverse to Γ-orbits for all $x \in \Lambda$. The rest of the argument proceeds by an upward induction on submaximal isotropy type. We omit the routine and straightforward details. □

Let $\mathcal{D}$ denote the family of all slices for Λ of the form $D(x, s)$, $s \in (0, c_k d_k(x)]$. For each $D(x, s) \in \mathcal{D}$, we set $H = H_D = \Gamma_x$. Thus, H_D is the maximal compact subgroup of Γ leaving D invariant.

6.3. Pre-Markov partitions

DEFINITION 6.3.1. A finite set $\mathcal{P}$ of rectangles is called a *pre-Markov partition* for $f : \Lambda \to \Lambda$ if $\cup_{R \in \mathcal{P}} R = \Lambda$ and, for all $R, S \in \mathcal{P}$, we have

(a) $R^\circ \cap S^\circ = \emptyset$, $R \neq S$.

(bs) $x \in R, f(x) \in S^\circ \Longrightarrow f(W^s(\Gamma x, R)) \subset W^s(\Gamma f(x), S)$.
(bu) $x \in R, f^{-1}(x) \in S^\circ \Longrightarrow f^{-1}(W^u(\Gamma x, R)) \subset W^u(\Gamma f^{-1}(x), S)$.

If the rectangles in $\mathcal{P}$ are all strict, we say that $\mathcal{P}$ is a *strict* pre-Markov partition.

DEFINITION 6.3.2. Let $\mathcal{P}$ be a pre-Markov partition. We define the *mesh* of $\mathcal{P}$ by $\mathrm{mesh}(\mathcal{P}) = \max\{\delta(R) \mid R \in \mathcal{P}\}$.

PROPOSITION 6.3.3. *Every basic set admits strict pre-Markov partitions $\mathcal{R}$ of arbitrarily small mesh . Furthermore, if mesh($\mathcal{R}$) is sufficient small, we can require that for each $R \in \mathcal{R}$, there exists a slice $D \in \mathcal{D}$ through a point $x \in R$, $\rho(R) = (\Gamma_x)$, such that $R \subset \Gamma(D)$.*

PROOF. The proof is similar to that of Theorem 4.2.6. Using an induction over isotropy type, we construct a cover of Λ by interiors of strict rectangles. We then follow a straightforward generalization of Bowen's original proof [**8**]. Alternatively, we can use a method based on shadowing (cf Remark 4.2.7). Specifically, for $\gamma > 0$ sufficiently small, choose a finite γ-dense subset $\mathcal{B}$ of Λ/Γ such that $\mathcal{B}$ intersects the quotient of each orbit stratum in a γ-dense set. Let $\Sigma \subset \mathcal{B}^{\mathbb{Z}}$ denote the associated subshift of finite type [**66**, 9.6]. Although we do not have uniqueness of shadowing in Λ/Γ, Proposition 5.7.3 does naturally associate a point $\theta(z)$ to each pseudo-orbit in Σ. Lifting back to Λ we obtain a cover of Λ/Γ by rectangles. Just as in [**66**, 9.6], this cover can then be refined to a pre-Markov partition of Λ. □

EXAMPLE 6.3.4. Let $\sigma : 3^{\mathbb{Z}} \to 3^{\mathbb{Z}}$ be the $\mathbb{Z}_2$-invariant full shift on three symbols (see Example 4.2.12). Regard $\mathbb{Z}_2$ as a subgroup of S^1 and form the twisted product $\Lambda = 3^{\mathbb{Z}} \times_{\mathbb{Z}_2} S^1$. Set $f = \sigma_\beta$, where $\beta : 3^{\mathbb{Z}} \to S^1$ is continuous. If we take the 'standard' three rectangle Markov partition for $3^{\mathbb{Z}}$, then the corresponding pre-Markov partition of Λ consists of just two rectangles, say R_0, R_1. In this case it is easy to see that $\cap_{i\in\mathbb{Z}} f^i(R_{j_i})$ will consist of more than one S^1-orbit (unless, for example, $j_i = 0$ for all $i \in \mathbb{Z}$). If instead we use the finer Markov partition for $3^{\mathbb{Z}}$ consisting of 9 rectangles, the corresponding pre-Markov partition for Λ consists of five rectangles. Using Example 4.2.12, we again find that infinite intersections typically contain more than one S^1-orbit. ♡

Just as for rectangles when there is no symmetry, we may define $\partial^s R$, $\partial^u R$ for a strict rectangle R. Then $\partial R = \partial^s R \cup \partial^u R$ and $\partial^s R, \partial^u R$

are closed Γ-invariant subsets of R. If $\mathcal{P}$ is a (strict) pre-Markov partition of Λ, define $\partial^s\mathcal{P} = \cup_{R\in\mathcal{P}}\partial^s R$, $\partial^u\mathcal{P} = \cup_{R\in\mathcal{P}}\partial^u R$. It is straightforward to verify that

$$f(\partial^s\mathcal{P}) \subset \partial^s\mathcal{P},\ \ f^{-1}(\partial^u\mathcal{P}) \subset \partial^u\mathcal{P}. \tag{6.1}$$

6.4. Proper and admissible rectangles

In this section we give our extension of Γ-regular partitions to connected compact Lie groups (we leave the extension to general compact Lie groups to the reader). First, we need some preliminaries.

Throughout this section, $A = A_\rho$ will always denote an open ρ-disk neighborhood of the identity in Γ, such that for all $x \in \Lambda$, and all $\gamma \in \Gamma_x \setminus \{e\}$, $\gamma A \cap A = \emptyset$. Given $x \in \Lambda$, choose a slice $D \in \mathcal{D}$. Suppose that $x \in D$ and for some $\eta > 0$, $W^{ss}_\eta(x), W^{uu}_\eta(x) \subset \Gamma D$. For small enough $\eta > 0$ (independent of D, x) we have $W^{ss}_\eta(x), W^{uu}_\eta(x) \subset A(D)$. We define

$$\tilde{W}^s_\eta(x) = A(W^{ss}_\eta(x)) \cap D,\ \ \tilde{W}^u_\eta(x) = A(W^{uu}_\eta(x)) \cap D.$$

The submanifold $\tilde{W}^s_\eta(x)$ is Γ_x-equivariantly diffeomorphic to $W^{ss}_\eta(x)$. (A formal proof may be given using a Γ_x-equivariant local section $\xi : W \subset \Gamma/\Gamma_x \to \Gamma$ – see [**25**, §3].)

DEFINITION 6.4.1. Let $D \in \mathcal{D}$. A compact set $S \subset D$ is a *proper D-rectangle* if

(a) $S = \overline{S^\circ}$ (interior relative to D).
(b) There exists $\eta > 0$ such that for all $x, y \in S$, $\tilde{W}^u_\eta(x) \cap \tilde{W}^s_\eta(y)$ consists of a single point, lying in S.
(c) If $\gamma \in H_D$, then either $\gamma S^\circ = S^\circ$ or $\gamma S^\circ \cap S^\circ = \emptyset$.

In the sequel, we set $\tilde{W}^u_\eta(x) \cap \tilde{W}^s_\eta(y) = \langle x, y\rangle$, $x, y \in S$ and note that $\langle\ ,\ \rangle : S \times S \to S$ will be continuous.

DEFINITION 6.4.2. A rectangle $R \subset \Lambda$ is a *proper rectangle* if we can choose $x \in R$, with $\rho(R) = (\Gamma_x)$, slice $D = D(x, r) \in \mathcal{D}$, and a proper D-rectangle S such that $R = \Gamma S$.

REMARKS 6.4.3. (1) If $R = \Gamma(S)$ is a proper rectangle, then there exists a Γ-invariant subset $\hat{R} \subset U_\Delta$ of R^2 such that (a) If $x, y \in \hat{R}$, then $[\![x, y], [x, y]\!] \in R$, and (b) given $x, y \in R^2$ there exists $\gamma \in \Gamma$ such that $x, \gamma y \in \hat{R}$. For example, $\hat{R} = \Gamma S^2 \subset R^2$. (2) Every strict rectangle is proper but the converse is false. A simple example may be based on Example 6.3.4 and is constructed by taking the twisted product $\Lambda = 3^{\mathbb{Z}} \times_{\mathbb{Z}_2} S^1$ and a slice at the point $(\bar{0}, e)$, which we may identify with (a subset of) $\mathbb{R}^2$. Referring to Figure 1, we take R to be the S^1-orbit of S_1 (or S_2). Since S_1 is a proper $\mathbb{R}^2$-rectangle and $S_1^\circ \cap S_2^\circ = \emptyset$, R is proper. However, R is certainly not strict. $\diamondsuit$

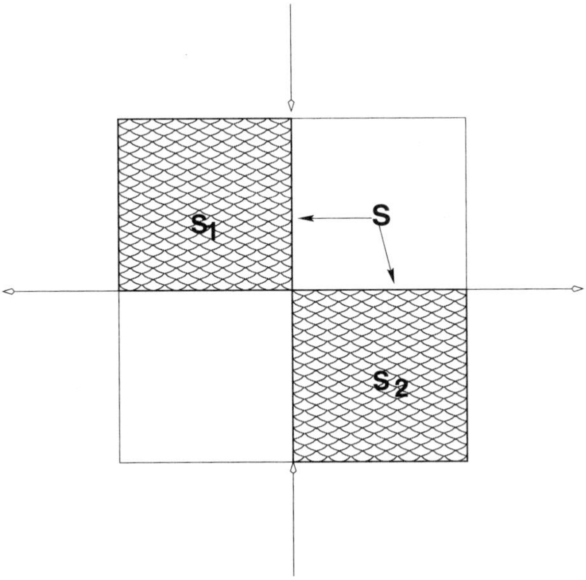

FIGURE 1. A proper rectangle which is not strict

We omit the straightforward proof of the following lemma.

LEMMA 6.4.4. *Every rectangle R can be written uniquely as a countable union $\cup_{\alpha \in I} R_\alpha$ of proper rectangles R_α satisfying*

(r1) $R_\alpha^\circ \cap R_\beta^\circ = \emptyset$, $\alpha \neq \beta$.

(r2) $\cup_{\alpha \in I} R_\alpha^\circ = R^\circ$.

Let S be a proper D-rectangle and set $H = H_D$. Then $\Gamma(S)$ is a proper rectangle and $\Gamma(S) \cap D = H(S)$ is a union of at most $|H|$ proper D-rectangles. Conversely, if R is a proper rectangle contained in $\Gamma(D)$, then $R \cap D$ will be an H-orbit of at most $|H|$ proper D-rectangles.

If S is a proper D-rectangle, define $\tilde{W}^s(S, x) = \tilde{W}_\eta^s(x) \cap S$, where as usual $0 < \eta(S) \ll \delta$. We similarly define $\tilde{W}^u(S, x)$. In general, $\Gamma\tilde{W}^s(S, x)$ will be a *proper* subset of $W^s(\Gamma x, \Gamma S)$. We define $\partial^s S$, $\partial^u S$ in the usual way. In particular, we have $\partial S = \partial^s S \cup \partial^u S$. Without further conditions on S, it is easy to find examples of proper D-rectangles for which $\partial\,\Gamma(S)$ is a proper subset of $\Gamma(\partial^s S \cup \partial^u S)$.

DEFINITION 6.4.5. Let $\varepsilon > 0$. A proper D-rectangle S is said to be *ε-admissible* if for all $x \in S^\circ$, and non-identity elements γ of $H = H_D$ we have

(1s) $\gamma \tilde{W}^s_\varepsilon(x) \cap S = \emptyset$,
(1u) $\gamma \tilde{W}^u_\varepsilon(x) \cap S = \emptyset$.
(2) $\partial S \cap \partial\gamma S \subset \Lambda_S$.

Generally, we say S is *admissible* if S is ε-admissible for some $\varepsilon > 0$.

A proper rectangle R is (ε-) admissible if we can write $R = \Gamma S$, for some (ε-) admissible proper D-rectangle S.

If S is an ε-admissible proper D-rectangle, we let $S(\varepsilon, s)$ (resp $S(\varepsilon, u)$) be the set of points $x \in S$ for which (1s) (resp (1u)) is satisfied. We similarly define $R(\varepsilon, s)$, $R(\varepsilon, u)$ for admissible proper rectangles. Obviously, $R(\varepsilon, s) \cap R(\varepsilon, u) \supset R^\circ$.

REMARK 6.4.6. It may be shown that the definitions of $R(\varepsilon, s)$, $R(\varepsilon, u)$ are independent of choice of proper D-rectangle or slice D. ♢

LEMMA 6.4.7. *If the proper D-rectangle S is admissible then for all $x \in S^\circ$ we have*

(1) $\gamma \tilde{W}^s(x, S) \cap \tilde{W}^s(x, S) = \emptyset$, $\gamma \tilde{W}^u(x, S) \cap \tilde{W}^u(x, S) = \emptyset$, $\gamma \neq e_H$.
(2) $\tilde{W}^s_\varepsilon(x) \cap S = \tilde{W}^s_\varepsilon(x) \cap H_D S$, $\tilde{W}^u_\varepsilon(x) \cap S = \tilde{W}^u_\varepsilon(x) \cap H_D S$.
(3) $\Gamma \tilde{W}^s(x, S) = W^s(\Gamma x, \Gamma S)$, $\Gamma \tilde{W}^u(x, S) = W^u(\Gamma x, \Gamma S)$.

Furthermore,

$$\partial \Gamma S = \Gamma \partial S = \Gamma \partial^s S \cup \Gamma \partial^u S.$$

PROOF. Since $\gamma x \notin S^\circ$, (1,2,3) follow trivially from the definition of admissibility. Since $\partial S \cap \partial \gamma S \subset \Lambda_S$, it follows that no point of ∂S can be an interior point of HS, proving the final statement. □

REMARK 6.4.8. If $R = \Gamma S$ is an admissible rectangle, with S an admissible D-rectangle, we may define $\partial^s R = \Gamma \partial^s S$, $\partial^u R = \Gamma \partial^u S$. It follows from Lemma 6.4.7 that $\partial R = \partial^s R \cup \partial^u R$. ♢

6.5. Γ-regular Markov partitions

DEFINITION 6.5.1. Let $\varepsilon > 0$. An *(ε-admissible) Γ-regular Markov partition* $\mathcal{R}$ consists of a finite set of proper admissible rectangles such that

(1) $\cup_{R \in \mathcal{R}} R = \Lambda$.
(2) $R^\circ \cap S^\circ = \emptyset$, $R, S \in \mathcal{R}$, $R \neq S$.
(3s) If $x \in R$, $f(x) \in S^\circ$, then $f(W^s(\Gamma x, R)) \subset W^s(f(x), S)$.
(3u) If $x \in R$, $f^{-1}(x) \in S^\circ$, then $f^{-1}(W^u(\Gamma x, R)) \subset W^u(f^{-1}(x), S)$.
(4) $\exists \varepsilon > 0$ for which

$$\cup_{R \in \mathcal{R}} R(\varepsilon, s) \supset \Lambda \setminus W^s_\varepsilon(\Lambda_S), \ \cup_{R \in \mathcal{R}} R(\varepsilon, u) \supset \Lambda \setminus W^u_\varepsilon(\Lambda_S).$$

REMARK 6.5.2. In the sequel, when we refer to a Γ-regular Markov partition $\mathcal{R}$, we mean that $\mathcal{R}$ is ε-admissible for some $\varepsilon > 0$. ♢

EXAMPLE 6.5.3. Let $\mathcal{S}$ be the $\mathbb{Z}_2$-regular Markov partition for the full $\mathbb{Z}_2$-shift $\sigma : \mathbf{3}^{\mathbb{Z}} \to \mathbf{3}^{\mathbb{Z}} \to$ that was constructed in Example 4.2.26 (see Figure 1(B)). Regard $\mathbb{Z}_2$ as the subgroup of $\mathbb{K}$ generated by a rotation through π and form the twisted product $\Lambda = \mathbf{3}^{\mathbb{Z}} \times_{\mathbb{Z}_2} \mathbb{K}$. Let $f : \mathbf{3}^{\mathbb{Z}} \to \mathbb{K}$ be a continuous $\mathbb{Z}_2$-invariant cocycle and let σ_f denote the corresponding $\mathbb{K}$-equivariant homeomorphism induced on Λ. If we define $\mathcal{R} = \{\mathbb{K}S \mid S \in \mathcal{S}\}$, then $\mathcal{R}$ is a $\mathbb{K}$-regular Markov partition on Λ. ♡

Suppose $\mathcal{R}$ is a Γ-regular Markov partition. We define $\partial^s\mathcal{R} = \cup_{R\in\mathcal{R}}\partial^s R$, $\partial^u\mathcal{R} = \cup_{R\in\mathcal{R}}\partial^u R$. It is straightforward to verify that

$$f(\partial^s\mathcal{R}) \subset \partial^s\mathcal{R}, \quad f^{-1}(\partial^u\mathcal{R}) \subset \partial^u\mathcal{R}. \tag{6.2}$$

Let $R \in \mathcal{R}$. Given $x \in R$, we may choose a slice $D = D(R; \bar{x})$ such that $x \in D$ and $R \cap D = R_D$ is a proper D-rectangle. Thus, we may take D to be a slice through a point $\bar{x}$ of R such that $(\Gamma_{\bar{x}}) = \rho(R)$, $x \in D$ and $\Gamma D \supset R$.

Suppose $x \in R$, $f(x) \in S^\circ$. It follows from (6.2) that $x \in R^\circ$. Choose associated slices D for R, E for S so that $x \in D$, $f(x) \in E$. Since $D \cap R$ consists of $|H_D|$ proper D-rectangles, we may choose a (unique) proper D-rectangle $R_\bullet$ such that $R = \Gamma R_\bullet$ and $x \in R_\bullet^\circ$. Similarly, we may choose a unique proper E-rectangle $S_\bullet$ such that $f(x) \in S_\bullet^\circ$. Using the admissibility of the rectangles R, S together with properties (3s,u) for a Γ-regular Markov partition, it is straightforward to verify that, provided mesh$(\mathcal{R})$ is sufficiently small (compared with the diameter of A), we have

$$f(R_\bullet) \cap S \subset A(S_\bullet). \tag{6.3}$$

We have an analogous result with f^{-1} replacing f.

REMARK 6.5.4. The relation (6.3) corresponds to (b) of Definition 4.2.15 as it implies that $f(R_\bullet^\circ) \cap A(\gamma S_\bullet^\circ) = \emptyset$ for all non-identity elements γ of H_E. ♢

Continuing with our assumption that $\mathcal{R}$ is a Γ-regular Markov partition of Λ, let $p : \Lambda \to \Lambda/\Gamma$ denote the orbit map. As we did for finite groups, we may define a partition $\tilde{\mathcal{R}}$ of Λ/Γ by

$$\tilde{\mathcal{R}} = \{p(R) \mid R \in \mathcal{R}\}.$$

Just as we did for finite groups, we define $\tilde{R}^\sigma = \tilde{R} \setminus p(\Lambda_S)$. Note that since $\Lambda_S \cap R \subset \partial R$, $(\tilde{R}^\sigma)^\circ = \tilde{R}^\circ$.

LEMMA 6.5.5. *Let $\tilde{R} \in \tilde{\mathcal{R}}$. For each $\alpha \in \tilde{R}^\sigma$, the sets $W^s(\alpha, \tilde{R}) = pW^s(x, R)$, $W^u(\alpha, \tilde{R}) = pW^u(x, R)$, are defined independently of the choice of $R \in \mathcal{R}$, such that $p(R) = \tilde{R}$, and $x \in R$ such that $p(x) = \alpha$.*

PROOF. Similar to that of Lemma 4.2.18, using the admissibility of R. □

LEMMA 6.5.6. *Let $R \in \mathcal{R}$. Let D be a slice through a point of R with isotropy $\rho(R)$ and suppose that $R \cap D$ consists of $k = |H_D|$ proper D-rectangles, say $R_1, \dots, R_k$. Then*

(1) *$p(R^\circ) = \tilde{R}^\circ$ and $\overline{\tilde{R}^\circ} = \tilde{R}$.*
(2) *$p(R_i) = \tilde{R}$, $1 \leq i \leq k$.*
(3) *There is a natural continuous map $[\![\, , \,]\!] : \tilde{R}^\sigma \times \tilde{R}^\sigma \to \tilde{R}$ such that if $x, y \in \tilde{R}^\sigma$ and for some (any) i we choose $x', y' \in R_i$ such that $x = p(x')$, $y = p(y')$, then $[\![p(x), p(y)]\!] = p\langle x', y' \rangle$.*

PROOF. Statements (1,2) follow immediately from the definitions. That the definition of $[\![\, , \,]\!]$ on $(\tilde{R}^\sigma)^2$ is independent of choices follows just as in the case when Γ was finite – Lemma 4.2.19. □

In the sequel, we refer to the sets $\tilde{R}$ as *rectangles*.

Associated to $\mathcal{R} = \{R_i \mid i \in I\}$ we may define a 01-matrix A by requiring that the entry $a_{ij} = 1$ if there exist $x \in R_i$ with $f(x) \in R_j^\circ$. The matrix A is equal to the corresponding 01-matrix $\tilde{A}$ associated to $\tilde{\mathcal{R}}$.

THEOREM 6.5.7. *Let $\mathcal{R}$ be a Γ-regular Markov partition for Λ and A be the associated $m \times m$ 01-matrix. Provided that mesh($\mathcal{R}$) is sufficiently small, the map $\pi : \Sigma_A \to \Lambda/\Gamma$ defined by*

$$p((x_i)) = \cap_{n \in \mathbb{Z}} \tilde{f}^{-n}(p(R_{x_n})), \quad ((x_i) \in \Sigma_A),$$

is well defined, Hölder continuous and surjective. Furthermore,

(a) *$\tilde{f}\pi = \pi\sigma$.*
(b) *$\#\pi^{-1}(x) \leq m^2$, $(x \in \Lambda/\Gamma)$.*
(c) *$\#\pi^{-1}(x) = 1$ for x lying in a residual subset of Λ.*

PROOF. Let $\mathbf{y}, \mathbf{z} \in \cap_{n \in \mathbb{Z}} \tilde{f}^{-n}(\tilde{R}_{x_i})$, $((x_i) \in \Sigma_A)$. We prove that $\mathbf{y} = \mathbf{z}$ using equivariant shadowing. Following the statement of Theorem 5.7.4, we suppose that $\rho, \beta > 0$ are chosen sufficiently small so that uniqueness of equivariant shadowing holds. In particular, A will be a ρ-disk neighborhood of the identity in Γ. Set $R_{x_0} = R_0$. Let D_0 be a slice through a maximal isotropy point of R_0 such that $\Gamma D_0 \supset R_0$ and $R_0 \cap D_0$ consists of $|H_{D_0}| = k$ proper D_0-rectangles, $R_0^1, \dots, R_0^k$. Choose $y_0, z_0 \in R_0^1$ such that $p(y_0) = \mathbf{y}$, $p(x_0) = \mathbf{z}$. Set $y_1 = f(y_0)$, $z_1 = f(z_0)$. Choose a slice D_1 through a maximal isotropy point of $R_{x_1} = R_1$ such that $\Gamma D_1 \supset R_1$, $y_1 \in D_1$, $R_1 \cap D_1$ consists of $|H_{D_1}|$ proper D_1-rectangles, $\{R_1^1, \dots, R_1^m\}$. There exists a unique

j, $1 \le j \le m$, for which there exist points of R_0^1 mapped to interior points of R_1^j. It follows by the (uniform) continuity of f that, if $\mathrm{mesh}(\mathcal{R})$ is sufficiently small, then $f(z_1) \in A(R_1^j)$. Now choose $\gamma_1 \in A$ such that $\gamma z_1 = \hat{z}_1 \in R_1^j$. Proceeding inductively, we construct sequences $(\hat{z}_i) \subset \Lambda$ and $(\gamma_i) \subset A$, $\gamma_0 = e$, such that for all $i \in \mathbb{Z}$ we have

$$\hat{z}_i \in \Gamma f^i(z_0), \quad d(\hat{z}_i, y_i) < \mathrm{mesh}(\mathcal{R}), \quad d(\hat{z}_i, \Pi_{j=0}^i \gamma_i z_i) = 0.$$

It follows by uniqueness of equivariant shadowing, that if $\mathrm{mesh}(\mathcal{R})$ is sufficiently small, the pseudo-orbit $(\hat{z}_i)$ is uniquely equivariantly shadowed by the f-orbit of a unique (mod Γ) point. But, by our construction, both (y_i) and (z_i) equivariantly shadow $(\hat{z}_i)$. Hence y_0 and z_0 lie on the same Γ-orbit and therefore $\mathbf{y} = \mathbf{z}$. The remaining parts of the proof are straightforward and involve standard methods. In particular, Hölder continuity follows from Hölder continuity properties of the stable and unstable foliations (that is, the Hölder continuity of the maps $[\![\ ,\ ]\!], [\ ,\ ]\!]$). □

6.6. Construction of Γ-regular Markov partitions

THEOREM 6.6.1. *Let Λ be a basic set for the Γ-equivariant diffeomorphism $f : M \to M$. Assume that Λ_Π consisting of points of trivial isotropy is open and dense in Λ. Then Λ admits Γ-regular Markov partitions $\mathcal{R}$ of arbitrarily small mesh.*

PROOF. Our proof is fairly similar to that of Theorem 4.4.1. We start with a pre-Markov partition and refine it to obtain a Γ-regular partition. The main difference with the proof of Theorem 4.4.1 is that when we do the refinement we have to work with proper D-rectangles rather than just rectangles. This is an unavoidable problem caused by the fact that since the group is continuous, rectangles are Γ-invariant sets and so Γ can never act freely on the set of rectangles. Given $\eta > 0$, it follows from Proposition 6.3.3 that Λ has a pre-Markov partition $\mathcal{P}$ with $\mathrm{mesh}(\mathcal{P}) < \eta$. We define

$$\mathcal{P}^1 = \{R \in \mathcal{P} \mid R \cap \Lambda_S \neq \emptyset\}, \quad \mathcal{P}^2 = \mathcal{P} \setminus \mathcal{P}^1.$$

For sufficiently small $\eta > 0$, $\mathcal{P}^2 \neq \emptyset$. Let $\mathbf{R}^i = \cup_{R \in \mathcal{P}^i} R$, $i = 1, 2$, and define $\mathbf{U} = \mathbf{R}^1 \setminus \cup_{x \in \Lambda_S} W^{ss}_{\mathrm{loc}}(x)$. It follows from transverse hyperbolicity and f-invariance of Λ_S that $\mathbf{U}$ is an open and dense Γ-invariant subset of $\mathbf{R}^1$. For sufficiently small $\eta > 0$, it is true that for any $x \in \mathbf{U}$, there exists a smallest $N = N(x) > 0$ such that $f^i(x) \in U$, $0 \le i < N$ and $f^N(x) \notin \mathbf{R}^1$. Exactly as in the proof of Theorem 4.4.1, we define for all $R \in \mathcal{P}^1$, $T \in \mathcal{P}^2$, and $n \in \mathbb{N}$, open subsets $R(T)^\circ_n$, $R(T)^{\circ\circ}$ of R°. We let $R(T)_n$ and $R(T)$ denote the corresponding closures. These sets are Γ-invariant subsets of R and $R(T)_n$, $R(T)$ are proper sub-rectangles of R.

Associated to each $R \in \mathbf{P}^1$, choose a slice $D = D(R)$ through a point of maximal isotropy of R such that $\Gamma(D) \supset R$. Let $k = k_R = |\rho(R)| = |H_D|$ and set $R_D = D \cap R = R_D$, $R(T)_n \cap D = R_D(T)_n$.

Let $T \in \mathcal{P}^2$ and suppose that $R(T)_1^\circ \neq \emptyset$. Observe that $f(R_D)$ intersects T in k disjoint pieces $T_1, \ldots, T_\ell$, permuted by H_D. Define $R_D(T)_1^i = f^{-1}(T_i) \cap R_D$. The sets $R_D(T)_1^i$ are obviously proper D-rectangles and $R_D(T)_1 = \cup_i R_D(T)_1^i$. We carry out this construction for all $R \in \mathbf{P}^1$ and $T \in \mathbf{P}^2$.

We now extend the previous construction to obtain a decomposition of each proper D-rectangle $R_D(T)_n$ into a union of k proper D-rectangles $R_D(T)_n^i$, permuted by H_D. We carry out this construction so that the correct incidence conditions on images of rectangles by f are preserved. Our proof is inductive. We suppose that at stage $n > 1$ we have for each $R \in \mathcal{P}^1$, $T \in \mathcal{P}^2$ obtained an $H(D)$-invariant decomposition of $R_D(T)_n$ into mutually disjoint proper D-rectangles $R_D(T)_n^i$, $1 \leq i \leq k$ such that the following conditions hold.

(a) $\Gamma R_D(T)_n^i = \cup_{j=1}^n R(T)_n$, $1 \leq i \leq k$.
(b) $R_D(T)_{n-1}^i \subset R_D(T)_n^i$, $1 \leq i \leq k$.
(c) If $R_D(T)_{n-1}^i \neq R_D(T)_n^i$ and there exists

$$x \in \partial^s R_D(T)_{n-1}^i \cap H_D \overline{R_D(T)_n^i \setminus R_D(T)_{n-1}^i},$$

then $x \notin \partial^s R_D(T)_n^i$.
(d) If $x \in R_D(T)_n^i$ and $f^n(x) \in T^\circ$ then, for a unique $R' \in \mathcal{P}^1$,

$$\Gamma f(\tilde{W}^u(x, R_D(T)_n^i)) \supset W^u(f(x), \Gamma R'(T)_{n-1}).$$

Condition (c) is included to allow for the situation when there are adjacent sections of R mapped to T. The condition implies that if $x \in \partial^s R_D(T)_n^i$, then $\gamma \tilde{W}_\varepsilon^s(x, R_D(T)_n^i) \cap R_D(T)_n^i = \emptyset$, $\gamma \in H_D$, $\gamma \neq Id$. Otherwise put, we minimize the (stable) boundary of $R_D(T)_n^i$.

The construction of $R_D(T)_{n+1}^i$, given the existence of $R'_D(T')_n^i$ for all R', T', is routine and uses the incidence properties of the pre-Markov partition $\mathcal{P}$. Set $R(T)^i = \overline{\cup_{n \geq 1} R_D(T)_n^i}$, $1 \leq i \leq k$. Each $R(T)_i$ is a proper D-rectangle and H_D acts transitively on $\{R(T)^1, \ldots, R(T)^k\}$.

We repeat the previous construction using f^{-1} in place of f. Thus for $R \in \mathcal{P}^1$, $S \in \mathcal{P}^2$, we construct a set $\{\bar{R}^i(S) | 1 \leq i \leq k\}$ of proper D-rectangles permuted by H_D.

Suppose $R \in \mathcal{P}^1$, $S, T \in \mathcal{P}^2$. If $R(T)^i \cap \bar{R}(S)^j \neq \emptyset$ for some i, j, then it follows easily by equivariance that $R(S,T)^{i,j} = R(T)^i \cap \bar{R}(S)^j \neq$

$\emptyset$ for all i, j. It follows from our construction that each $R(S,T)^{i,j}$ is an admissible proper D-rectangle. Re-indexing, we write the set $\{R(S,T)^{i,j} \mid 1 \leq i,j \leq k\}$ as a union of k sets $\{R(S,T)^i_j | 1 \leq j \leq k\}$, $1 \leq i \leq k$, where H_D acts transitively on each $\{R(S,T)^i_j | 1 \leq j \leq k\}$. Finally, we define $R(S,T)^i = \Gamma R(S,T)^i_j$, noting that $R(S,T)^i$ is independent of the choice of j. The sets $R(S,T)^i$ are admissible rectangles. Finally, we define the required Γ-regular Markov partition $\mathcal{R}$ of Λ by replacing $\mathcal{P}^1$ by the set $\mathcal{P}^\star$ of all (non-empty) admissible rectangles $R(S,T)^i$, $R \in \mathcal{P}^1$, $S, T \in \mathcal{P}^2$. Incidence conditions between rectangles in $\mathcal{R}$ hold either by construction of $\mathcal{P}^\star$ or because they held for rectangles in $\mathcal{P}^2$. □

REMARK 6.6.2. Note that in the final step of the proof of Theorem 6.6.1 it is not necessary to refine rectangles in $\mathcal{P}^2$. Indeed, that step would not have been necessary in the proof of Theorem 4.4.1 had we only wanted a Markov partition on Λ yielding a symbolic dynamics on Λ/Γ. The point being that the condition $f(S^\circ) \cap R^\circ \neq \emptyset \implies f(S^\circ) \cap \gamma R^\circ = \emptyset$, $\gamma \neq e$ is only really used to obtain a symbolic dynamics on Λ/Γ when $R \cap \Lambda_S \neq \emptyset$. Note, however, that by refining rectangles in $\mathcal{P}^2$ which are close to Λ_S, and taking the mesh($\mathcal{R}$) sufficiently small, we can arrange that all rectangles in $\mathcal{R}$ are ε-admissible where $\varepsilon >$ mesh($\mathcal{R}$) and that condition (4) of Definition 6.5.1 holds with this values of ε. ♢

Part 2

Stable Ergodicity

CHAPTER 7

Preliminaries

In this chapter we briefly review some standard theory not included in Part I. Throughout we follow the (standard) notational conventions of Part I.

7.1. Metrics

Suppose that M is a smooth compact connected riemannian Γ-manifold with associated Γ-invariant metric d. Let Λ be a closed Γ-invariant subset of M. We define a metric $\bar{d}$ on the orbit space Λ/Γ by $\bar{d}(x,y) = d(p^{-1}(x), p^{-1}(y))$, where $p : \Lambda \to \Lambda/\Gamma$ denotes the orbit map.

REMARK 7.1.1. It follows from the theorem of Mostow-Palais [**49, 54**] that M can be equivariantly embedded in a finite dimensional orthogonal representation, say $(\mathbb{R}^n, \Gamma)$. Let $p_1, \ldots, p_\ell$ be a minimal set of homogeneous generators for the $\mathbb{R}$-algebra of invariant polynomials on $\mathbb{R}^n$ and set $P = (p_1, \ldots, p_\ell)$. The orbit space $\mathbb{R}^n/\Gamma$ is homeomorphic to $P(\mathbb{R}^n)$ and, by Schwarz' theorem, the homeomorphism is a diffeomorphism with respect to the natural smooth structures on $\mathbb{R}^n/\Gamma$ and $P(\mathbb{R}^n) \subset \mathbb{R}^\ell$ [**70**]. In particular, the Euclidean metric on $\mathbb{R}^\ell$ induces a metric on $\mathbb{R}^n/\Gamma$. Let $\hat{d}$ denote the corresponding metric on Λ/Γ. While $\hat{d}$ is *not* equivalent to $\bar{d}$, it is not hard to show using inductive techniques similar to those, for example, in [**70**] that there exists $0 < \alpha < 1$, and constants c_1, c_2 such that for all $x, y \in \Lambda/\Gamma$ (or M/Γ), we have $c_1\hat{d}(x,y) \leq \bar{d}(x,y) \leq c_2\hat{d}(x,y)^\alpha$. In particular, functions that are Hölder with respect to $\bar{d}$ will be Hölder with respect to $\hat{d}$, though with a smaller Hölder exponent. ♢

7.2. The Haar lift

Let $s : \Lambda/\Gamma \to \Lambda$ be a section of the orbit map (we do not assume s is continuous). Suppose that μ is a Borel measure on Λ/Γ. The *Haar lift* of μ to Λ is the Borel measure ν defined on Borel subsets A of Λ by

$$\nu(A) = \int_{pA} [\int_\Gamma \chi_A(\eta s(x)) dh(\eta)] d\mu(x).$$

REMARK 7.2.1. The Haar lift is defined in [**43**] when the action of Γ is free and in [**51**] for more general Γ-actions. Since we are assuming that Λ is a compact subset of a Γ-manifold, it is easy to verify (using slice theory) that the Haar lift is defined independently of choice of section s and without any additional restrictions on the Γ-action. In particular, if we let Λ_S denote the set of singular orbits in Λ, then $\mu(\Lambda_S/\Gamma) = 0 \Longrightarrow \nu(\Lambda_S) = 0$. $\diamondsuit$

7.3. Isotropy and ergodicity

Let $F : M \to M$ be a smooth Γ-equivariant diffeomorphism. For the remainder of this work, we assume that $\Lambda \subset M$ is a basic set for F and set $\Phi = F|\Lambda$. In what follows, we shall generally assume that F is C^s, $s \geq 2$, although some of our results hold with F C^1. Let $\phi : \Lambda/\Gamma \to \Lambda/\Gamma$ be the homeomorphism induced by Φ on the orbit space.

Let ν be a Φ-invariant measure on Λ which is strictly positive on open subsets of Λ. A necessary condition for Φ to be ν-ergodic is that Φ has a dense orbit, say $(\Phi^n(z))$. The existence of a dense orbit implies restrictions on the action of Γ on Λ. Specifically, if we let H denote the isotropy group of z, then $H \subset \Gamma_x$ for all $x \in \Lambda$ and we have equality on an open and dense Γ-invariant subset $\Lambda_N \subset \Lambda$. Since Λ_N is Γ-invariant, H must be a normal subgroup of Γ. Since H acts trivially on Λ, it is no loss of generality to replace Γ by Γ/H and assume that Γ acts freely on Λ_N and that all isotropy groups are finite. In the sequel we always assume these conditions on the action of Γ on Λ. (See also the discussion in Remark 3.1.5.)

7.4. Γ-regular Markov partitions

Following 6.5, let $\mathcal{R}$ be a Γ-regular Markov partition on Λ. Let $\tilde{\mathcal{R}} = \{\tilde{R} \mid R \in \mathcal{R}\}$ denote the Markov partition induced on Λ/Γ. If $\tilde{R} \in \tilde{\mathcal{R}}$, we set $\tilde{R}^\sigma = \tilde{R} \setminus p(\Lambda_S)$, where Λ_S denotes the set of singular Γ-orbits in Λ. It follows from Lemma 6.5.5 that if $\tilde{R} \in \tilde{\mathcal{R}}$ then for each $\alpha \in \tilde{R}^\sigma$, the sets $W^s(\alpha, \tilde{R}) = pW^s(x, R)$, $W^u(\alpha, \tilde{R}) = pW^u(x, R)$, are defined independently of the choice of $R \in \mathcal{R}$, such that $p(R) = \tilde{R}$, and $x \in R$ such that $p(x) = \alpha$. It follows from Lemma 6.5.6 that for every $\tilde{R}^\sigma \in \tilde{\mathcal{R}}$, there is a natural continuous map $[\![\ ,\ ]\!] : \tilde{R}^\sigma \times \tilde{R}^\sigma \to \tilde{R}$ such that if $\alpha, \beta \in \tilde{R}^\sigma$ then $[\![\alpha, \beta]\!] = W^s(\alpha, \tilde{R}) \cap W^u(\beta, \tilde{R})$.

It follows from Theorem 6.5.7 that the Markov partition $\tilde{\mathcal{R}}$ determines a symbolic dynamics on Λ/Γ.

PROPOSITION 7.4.1. *Let Λ be a basic set for Φ. There exists a subshift of finite type $\sigma : \Omega \to \Omega$ and a Hölder continuous surjection $\pi : \Omega \to \Lambda/\Gamma$ such that*

(a) $\pi\sigma = \phi\pi$

(b) $\#\pi^{-1}(z) \leq N$ *for some constant* N

(c) $\#\pi^{-1}(z) = 1$ *for z lying in a residual subset of* Λ.

7.4.1. Holonomy transformations for basic sets. We continue to assume that $\mathcal{R}$ is a Γ-regular Markov partition on Λ.

LEMMA 7.4.2. *Let $R \in \mathcal{R}$ and suppose that $\alpha, \beta \subset R^\circ$ are Γ-orbits. We have well-defined Hölder continuous holonomy transformations $h^s_{\alpha,\beta} : W^s(\alpha, R) \to W^s(\beta, R)$, $h^u_{\alpha,\beta} : W^u(\alpha, R) \to W^u(\beta, R)$ defined by*

$$\begin{aligned} h^s_{\alpha,\beta}(x) &= W^{uu}(x, R) \cap W^s(\beta, R), \\ h^u_{\alpha,\beta}(x) &= W^{ss}(x, R) \cap W^u(\beta, R). \end{aligned}$$

PROOF. It suffices to observe that it follows from the admissibility of R (Definition 6.4.5) that $W^{uu}(x, R) \cap W^s(\beta, R)$, $W^{ss}(x, R) \cap W^u(\beta, R)$ both consist of exactly one point. □

7.5. Measures on the orbit space

Given a continuous mapping $T : X \to X$ of the compact metric space X, let $h(T)$ denote the topological entropy of T. If μ is a T-invariant measure on X, let $h_\mu(T)$ denote the measure theoretic entropy of T.

Let m denote the Parry measure (measure of maximal entropy) on (σ, Ω) and define the equilibrium measure $\mu = \pi \cdot m$ on Λ/Γ. Since $(\phi, \Lambda/\Gamma, \mu)$ and (σ, Ω, m) are isomorphic, it follows from the Variational Principle [**38**, Theorem 4.5.3] that

$$h_m(\sigma) = h(\sigma) = h(\phi) = h_\mu(\phi).$$

The measure μ is positive on open sets of Λ/Γ and nonatomic.

If we let ν denote the Haar lift of μ to Λ, then ν is positive on open sets of Λ and nonatomic.

LEMMA 7.5.1. $h(\Phi) = h(\phi)$. *In particular, ν is a measure of maximal entropy for (Φ, Λ). It is unique amongst Haar lifts to Λ.*

PROOF. Since $\phi : \Lambda/\Gamma \to \Lambda/\Gamma$ is a factor of $\Phi : \Lambda \to \Lambda$, $h(\phi) = h_\mu(\phi) \leq h_\nu(\Phi) \leq h(\Phi)$. Choose a finite set of (smooth) slices S_i, $1 \leq i \leq N$ for the action of Γ on Λ such that each $\Gamma(S_i)$ is compact and $\cup_{i=1}^N \Gamma(\text{Interior}(S_i)) = \Lambda$. It follows from the compactness of Λ and continuity of Φ that we can choose $\varepsilon_0 > 0$ such that if $d(x, y) \leq \varepsilon_0$, then there exist S_i, S_j such that $x, y \in \Gamma(S_i)$ and $\Phi(x), \Phi(y) \in \Gamma(S_j)$. Suppose $x, y \in \Gamma(S_i)$. There exist $\gamma, \gamma' \in \Gamma$ such that $\gamma x, \gamma' y \in S_i$. Let

$$\zeta(x, y) = \min_i \inf\{\rho(\gamma, \gamma') \mid \gamma x, \gamma' y \in S_i\},$$

where ρ is a left and right translation invariant metric on Γ. Note that $\zeta(x, y)$ is well-defined for all $x, y \in \Lambda$ such that $\bar{d}(\Gamma x, \Gamma y) < \varepsilon_0$. Since Φ is uniformly Lipschitz on Λ, we can choose $C > 0$ such that whenever $x, y \in \Lambda$, $\zeta(x, y) = 0$, and $\bar{d}(\Gamma\phi^i(x), \Gamma\phi^i(y)) < \varepsilon_0$, $i = 0, \dots, n$, then $\zeta(\Phi^n(x), \Phi^n(y)) \leq Cnd(x, y)$. In other words, providing the distance between successive ϕ-iterates of Γx, Γy remains small then the relative drift along Γ-orbits grows linearly in n.

Next, we briefly recall the definition of topological entropy formulated in terms of (n, ε)–spanning sets (for more details see [**38**, Chapter 3]). Suppose that $T : X \to X$ is a continuous map of the compact metric space (X, δ). For $n \in \mathbb{N}$, we define the metric δ_n on X by

$$\delta_n(x, y) = \max_{0 \leq i \leq n-1} \delta(T^i x, T^i y)$$

A subset $\mathcal{F}$ of X (n, ε)-spans X if $\mathcal{F}$ is ε-dense in X. Let $r_n(\varepsilon, T)$ denote the smallest cardinality of any (n, ε)-spanning set for X. The topological entropy of T is defined by

$$h(T) = \lim_{\varepsilon \to 0} \limsup_{n \to \infty} \frac{1}{n} \log r_n(\varepsilon, T)$$

Applying this to $\phi : \Lambda/\Gamma \to \Lambda/\Gamma$, we have

$$h(\phi) = \lim_{\varepsilon \to 0} \limsup_{n \to \infty} \frac{1}{n} \log r_n(\varepsilon, \phi).$$

Fix $\varepsilon \in (0, \varepsilon_0)$, $n \in \mathbb{N}$ and let $\mathcal{F}$ be an (n, ε)-spanning set for ϕ of minimum cardinality. Pick $\hat{\mathcal{F}} \subset \Lambda$ which is mapped 1:1 on $\mathcal{F}$ by π. Let $\dim(\Gamma) = g$. We may assume (see the proof of Lemma 5.1.3) that the equivariant riemannian metric d on M is chosen so that all Γ-orbits of points in Λ are totally geodesic g-dimensional submanifolds of M. It follows from standard dimension theory that we may choose a constant $K > 0$ such that for all $x \in \Lambda$, the smallest cardinality of any open cover of Γx by ε-disks is bounded above by $K\varepsilon^{-g}$. It remains to control drift along Γ-orbits. So suppose $x \in S_i$ and all points within ε of x lie in $\Gamma(S_i)$. Let $B_\varepsilon(x) = \{y \in S_i \mid d(x, y) \leq \varepsilon\}$. Choose an $\frac{\varepsilon}{Cn}$-dense subset $A(x)$ of $B_\varepsilon(x)$ which is of minimal cardinality. Suppose that $\dim(M) = m$. It follows by dimension theory that there exists $L > 0$, independent of x, ε, n, such that the cardinality of $A(x)$ is bounded by $L(\frac{\varepsilon}{Cn})^{m-g}$. It follows easily from our constructions that we can bound the cardinality of an (n, ε)-spanning set for Φ by $K\varepsilon^{-g}L(\frac{\varepsilon}{Cn})^{m-g}r_n(\varepsilon, \phi)$. Taking logs and limits, we see that $h(\Phi) \leq h(\phi)$ and the result follows. □

7.6. Spectral characterization of ergodicity and weak-mixing

We conclude this chapter with the following useful characterizations of ergodicity and weak-mixing due to Keynes and Newton [**40**].

PROPOSITION 7.6.1. *The dynamical system* (Φ, Λ, ν) *is not ergodic if and only if there exists a nontrivial irreducible unitary representation* $\Gamma \to U(d)$ *and a nonzero measurable* $w : \Lambda \to \mathbb{C}^d$ *such that*

$$\begin{aligned} w \circ \Phi &= w, \\ w\gamma &= \gamma w, \quad (\gamma \in \Gamma). \end{aligned}$$

PROPOSITION 7.6.2. *Suppose that* $(\phi, \Lambda/\Gamma, \mu)$ *is weak mixing and* (Φ, Λ, ν) *is ergodic. Then* (Φ, Λ, ν) *is not weak-mixing if and only if there exists a nontrivial irreducible representation* $\Gamma \to \mathbb{K}$, $c \in \mathbb{K}$ *and a nonzero measurable function* $w : \Lambda \to \mathbb{C}$ *satisfying*

$$w \circ \Phi = cw, \tag{7.1}$$

$$w\gamma = \gamma w, \quad (\gamma \in \Gamma). \tag{7.2}$$

REMARK 7.6.3. Although our formulation of these results is slightly different from that of Keynes and Newton, the proofs are the same. For example, by the spectral characterization of ergodicity, (Φ, Λ, ν) is ergodic if and only if the induced unitary operator $U : L^2(\Lambda, \nu) \to L^2(\Lambda, \nu)$ has $\lambda = 1$ as an eigenvalue with multiplicity one. Using harmonic analysis, we may decompose the unitary Γ-representation $L^2(\Lambda, \nu)$ as the orthogonal direct sum $L^2(\Lambda/\Gamma, \mu) \oplus_i V_i$, where the V_i are nontrivial irreducible unitary representations of Γ. If the conditions of Proposition 7.6.1 hold, we then construct an explicit nonzero element of the corresponding V_i, and hence of $L^2(\Lambda, \nu)$, which is a non-constant equivariant eigenfunction of U with eigenvalue equal to one. The converse is equally straightforward. ♢

CHAPTER 8

Livšic regularity and ergodic components

8.1. Livšic regularity

We have the following version of the Livšic regularity theorem [**45**, **46**], which plays a pivotal role in verifying generic stable ergodicity for (Φ, Λ, ν).

THEOREM 8.1.1 (Livšic Regularity Theorem [**45**]). *Let $\rho : \Gamma \to U(d)$ be a unitary representation of Γ. Let $w : \Lambda \to \mathbb{C}^d$ be a ν-measurable map satisfying*

$$w \circ \Phi = w, \tag{8.1}$$

$$w\gamma = \gamma w, \quad (\gamma \in \Gamma). \tag{8.2}$$

Then there exists a Hölder continuous $w' : \Lambda \to \mathbb{C}^d$ satisfying (8.1,8.2) and such that $w = w'$, ν ae.

PROOF. We start by showing that if w satisfies (8.1), then w is bounded on a full measure subset of Λ. Certainly, there exists $K > 0$ and a measurable set $B \subset \Lambda$ such that $\nu(B) > 0$ and $|w(x)| \leq K$ for all $x \in B$. Since w is Γ-equivariant (8.2), we may assume that B is a Γ-invariant subset of Λ. It follows from (8.1) that $|w| \leq K$ on the Φ- and Γ-invariant set $B^\infty = \cup_{n=-\infty}^{\infty} \Phi^n(B)$. Since w is Γ-equivariant, w induces a μ-measurable map $\tilde{w} : \Lambda/\Gamma \to \mathbb{C}^d/\Gamma$. Since $\tilde{w}$ is bounded on $p(B^\infty)$, it follows by the μ-ergodicity of ϕ, that $\mu(B^\infty) = 1$. Since B^∞ is Γ-invariant, it follows from the definition of ν that $\nu(B^\infty) = 1$.

Let $\Lambda_0 = \{x \in \Lambda \mid |w(x)| \leq K\}$. It follows from the previous paragraph that $\nu(\Lambda_0) = 1$ and Λ_0 is a Φ- and Γ-invariant subset of Λ. By Lusin's theorem [**67**, Theorem 2.24], we may choose a subset $L \subset \Lambda_0$ such that $\nu(L) > \frac{1}{2}$ and $w|L$ is uniformly continuous. Since w is Γ-equivariant, we may further assume that L is a Γ-invariant subset of Λ_0. Necessarily, $\mu(p(L)) = \nu(L)$.

Let $\mathcal{G}$ be the set of points $\alpha \in \Lambda_0/\Gamma$ for which

$$\lim_{N\to\infty} \frac{1}{N} \sum_{i=0}^{N-1} \chi_{pL}(\phi^{-i}\alpha) = \mu(pL).$$

Since $\phi : \Lambda/\Gamma \to \Lambda/\Gamma$ is μ-ergodic, $\mu(\mathcal{G}) = 1$. Define Λ_{00} to be the set of points $x \in \Lambda_0$ such that

$$\lim_{N\to\infty} \frac{1}{N} \sum_{i=0}^{N-1} \chi_L(\Phi^{-i}x) = \nu(L).$$

Since Φ covers ϕ and L is Γ-invariant, $\nu(\Lambda_{00}) = 1$.

Let $x, y \in \Lambda_{00}$ lie on the same local strong unstable manifold. Define $x_n = \Phi^{-n}(x)$, $y_n = \Phi^{-n}(y)$. It follows from (8.1) that for all $n \in \mathbb{Z}$ we have $w(x) = w(x_n)$, and so

$$|w(x) - w(y)| = |w(x_n) - w(y_n)|.$$

Since $x, y \in \Lambda_{00}$ and $\nu(L) > \frac{1}{2}$ we may choose subsequences $\{x_{n_k}\}$, $\{y_{n_k}\}$ such that $x_{n_k}, y_{n_k} \in L$ for all n_k. Since w is uniformly continuous on L it follows that $\lim_{k\to\infty} |w(x_{n_k}) - w(y_{n_k})| = 0$. Hence $w(x) = w(y)$ and so w is constant, certainly Hölder, on the local strongly unstable manifolds $W_\varepsilon^{uu}(z)$, $z \in \Lambda_{00}$. By equivariance, w extends to a Hölder continuous function on $W_\varepsilon^u(\Gamma z)$, for all $z \in \Lambda_{00}$. Replacing Φ by Φ^{-1}, shows that w is Hölder on $W_\varepsilon^s(\Gamma z)$, $z \in \Lambda_{00}$.

Our argument implies that given $\varepsilon > 0$, we can choose $\delta > 0$ such that if $x, y \in \Lambda_{00}$ and $y \in W_\delta^{ss}(x) \cup W_\delta^{uu}(x)$ then $w(x) = w(y)$.

It follows from 5.6 that we have a local product structure on Λ. It follows from Proposition 5.2.4 that for sufficiently small $\delta > 0$, we can choose an open neighborhood A of the identity in Γ such that if $U = \{(x, y) \in \Lambda^2 \mid d(x, y) \leq \delta\}$, then there are Hölder continuous maps $\rho : U \to A$, $[\ ,\ ] : U \to \Lambda$ characterized by

$$[x, y] \;=\; W_\delta^{ss}(\rho(x, y)x) \cap W_\delta^{uu}(y).$$

Since $w(\rho(x, y)x) = \rho(x, y)w(x)$, it follows that that if $x, y, [x, y] \in \Lambda_{00}$ then

$$\begin{aligned} |w(x) - w(y)| &= |w(x) - \rho(x, y)w(x)|, \\ &\leq \|I - \rho(x, y)\| |w(x)|, \\ &\leq d(x, y)^\alpha K, \end{aligned}$$

where K is the upper bound for w on Λ_0 and α is the Hölder exponent of ρ.

We claim there exists $\delta > 0$ such that for ν ae $x \in \Lambda_{00}$, $[x, y] \in \Lambda_{00}$ for ν ae $y \in \{z \in \Lambda_{00} \mid d(x, z) \leq \delta\}$ (absolute continuity). Assuming the claim, it follows that there is a unique Hölder continuous version of w on Λ.

For the absolute continuity property, we use a variation of the argument of Ruelle & Sullivan used to prove absolute continuity of foliations for hyperbolic basic sets [**69**, Theorem 1(d)], together with the existence of Γ-regular Markov partitions. Details are presented in the Appendix. □

REMARKS 8.1.2. (1) Since the absolute continuity property we use holds for Haar lifts of equilibrium states on Λ/Γ associated to Hölder continuous potentials, Theorem 8.1.1 holds for any measure ν which is the Haar lift of an equilibrium state defined by a Hölder continuous potential on Λ/Γ. See also the Appendix. (2) If instead of (8.1), we assume that $w \circ \Phi = cw$, where $c \in \mathbb{K}$, then the proof of Theorem 8.1.1 continues to apply, without change, to give a Hölder continuous solution $w' : \Lambda \to \mathbb{C}^d$, $w' = w$, ae. In particular, both Propositions 7.6.1, 7.6.2 have continuous versions (cf [**57**, Theorems 4.3, 4.4]). ♢

COROLLARY 8.1.3. *If Φ is topologically transitive (that is, Φ has a dense orbit) then (Φ, Λ, ν) is ergodic.*

PROOF. If $w : \Lambda \to \mathbb{C}^d$ is ν-measurable and satisfies (8.1,8.2) then the continuous version of w given by Theorem 8.1.1 is constant on a dense subset of Λ and hence is constant. Thus any w satisfying the conditions of Proposition 7.6.1 is trivial. □

COROLLARY 8.1.4. *If (Φ, Λ, ν) is ergodic, $\phi : \Lambda/\Gamma \to \Lambda/\Gamma$ is topologically mixing and Γ is semisimple then (Φ, Λ, ν) is Bernoulli.*

PROOF. If Γ is semisimple then any one-dimensional representation $\rho : \Gamma \to \mathbb{K}$ is trivial. Thus the equation (7.2) defines a function w which is constant on group orbits and hence drops to the orbit space Λ/Γ. Since $(\phi, \Lambda/\Gamma, \mu)$ is topologically mixing it is weak mixing (else, the subshift dynamics would not be topologically mixing). Hence any solution to (7.1) by a Γ invariant function is constant and $c = 1$. Furthermore since (Φ, Λ, ν) is measure-theoretically a skew-product, Rudolph's theorem [**68**] applies. That is, if (Φ, Λ, ν) is weak-mixing, then (Φ, Λ, ν) is Bernoulli. □

8.2. Structure of ergodic components

Following Brin [**12**, Theorem 1.1], we define an equivalence relation $\sim$ on Λ by $x \sim y$ if there exist $x_0, \ldots, x_n \in \Lambda$ such that

(a) $x_0 = x$, $x_n = y$.
(b) $x_{i+1} \in W^{rr}(x_i)$, where $r \in \{s, u\}$.

We denote the equivalence class of x by $C(x)$. Obviously, if $y \in C(x)$ then $C(x) \supset W^{ss}(y) \cup W^{uu}(y)$. We note the following properties of the sets $C(x)$.

$$y \in C(x) \implies x \in C(y), \tag{8.3}$$
$$\Phi^n(C(x)) = C(\Phi^n(x)), \quad (n \in \mathbb{Z}), \tag{8.4}$$
$$C(\gamma x) = \gamma C(x), \quad (\gamma \in \Gamma). \tag{8.5}$$

Define the closed set $Q(z) = \overline{\cup_{-\infty}^{\infty} \Phi^n(C(z))}$.

PROPOSITION 8.2.1 (cf [**11**, Theorem 1]). *For all $z \in \Lambda$, we have*

(a) $p(Q(z)) = \Lambda/\Gamma$.
(b) $Q(z)$ *is Φ-invariant.*
(c) $\gamma Q(z) = Q(\gamma z)$, *for all $\gamma \in \Gamma$.*
(d) *If $w \in Q(z)$ then $Q(z) = Q(w)$.*
(e) $\Phi | Q(z)$ *is topologically transitive.*

PROOF. Since Λ/Γ has a symbolic dynamics (Proposition 7.4.1), $p(W^{rr}(x))$ is a dense subset of Λ/Γ for all $x \in \Lambda$, $r \in \{u, s\}$, proving (a). Statements (b,c) are immediate from (8.5,8.5). In order to prove (d), suppose that $w \in Q(z)$. Then for $\delta > 0$, there exists $\gamma \in \Gamma$, $k \in \mathbb{Z}$ such that $d(\gamma, Id) < \delta$ and $\gamma w \in \Phi^k C(z)$. Since the sets $C(x)$ define a partition of Λ, $C(\Phi^k z) = C(\gamma w)$. Hence, using (8.4, 8.5), we see that

$$\begin{aligned} Q(z) &= \overline{\cup_{m \in \mathbb{Z}} \Phi^m C(\Phi^k z)}, \\ &= \overline{\cup_{m \in \mathbb{Z}} \Phi^m C(\gamma w)}, \\ &= \gamma \overline{\cup_{m \in \mathbb{Z}} \Phi^m C(w)}, \\ &= \gamma Q(w). \end{aligned}$$

Letting $\delta \to 0$, $\gamma \to e$ and so $Q(w) = Q(z)$.

It remains to prove (e). For this we use the original argument of Brin [**12**, Theorem 1.1]. Since the measure ν on Λ is positive on open subsets of Λ, it follows from the Poincaré recurrence theorem that recurrent points form a dense subset of Λ. Let $B_\delta(x)$ denote the open d-disk, center x, radius δ in Λ. It follows from the definition of topological transitivity that $\Phi | Q(z)$ is topologically transitive if given any $x, y \in Q(z)$, $\delta > 0$, then there exists $k = k(x, y, \delta)$ such that $B_\delta(y) \cap \Phi^k B_\delta(x) \neq \emptyset$.

This property follows from topological transitivity of the strong stable and unstable foliations of $Q(z)$ ([**12**, Definition 1.1]) and the two lemmas below which are proved in [**12**] when Λ is a partially hyperbolic manifold. Both proofs extend immediately to our context (for the second lemma, we use the density of recurrent points).

LEMMA 8.2.2 ([**12**, Lemma 2]). *Let* $x_{i+1} \in W^{ss}(x_i)$. *For each* $\varepsilon_1 > 0$, *there exists* $\varepsilon_2 > 0$ *such that if* $\Phi^k B_\delta(x) \cap B_{\varepsilon_2}(x_i) \neq \emptyset$ *for some* k, *then there exists* k' *such that* $\Phi^{k'} B_\delta(x) \cap B_{\varepsilon_1}(x_{i+1}) \neq \emptyset$.

LEMMA 8.2.3 ([**12**, Lemma 3]). *Let* $x_{i+1} \in W^{uu}(x_i)$. *Then for* $\varepsilon_1 > 0$ *there exists* $\varepsilon_2 > 0$ *such that if* $\Phi^k B_\delta(x) \cap B_{\varepsilon_2}(x_i) \neq \emptyset$ *for some* k, *then there exists* k' *such that* $\Phi^{k'} B_\delta(x) \cap B_{\varepsilon_1}(x_{i+1}) \neq \emptyset$.

This completes the proof of Proposition 8.2.1. □

COROLLARY 8.2.4. *The sets* $Q(z)$ *define a closed* Γ*-invariant partition of* Λ *by the ergodic components of* (Φ, Λ, ν). *In particular, if* $z \in \Lambda$ *and* $\gamma \in \Gamma$ *then either* $\gamma Q(z) = Q(z)$ *or* $\gamma Q(z) \cap Q(z) = \emptyset$.

REMARK 8.2.5. Let $\mathcal{Q}$ denote the partition of Λ into closed Φ-invariant sets given by Proposition 8.2.1. If $Q \in \mathcal{Q}$, then Q must contain points of every isotropy type occurring in Λ (since $p(Q) = \Lambda/\Gamma$). In particular, if we let J denote the closure of the subgroup of Γ generated by $\{\Gamma_x \mid x \in Q\}$, then Q is J-invariant. Note that J must always contain a subgroup of maximal isotropy type. Hence, when the action of Γ on Λ is not free – that is, Λ is not a principal extension over a locally maximal set – then the components Q of $\mathcal{Q}$ are always stabilized by a non-trivial subgroup of Γ. ◇

CHAPTER 9

Stable Ergodicity

Throughout this chapter we assume that Λ is a basic set for the C^s-diffeomorphism Φ, $s \geq 2$. Let $\phi : \Lambda/\Gamma \to \Lambda/\Gamma$ denote the map induced on the orbit space by Φ. Let $r \in (0, s]$. Following §2.2.2, we let $C^r_\Gamma(\Lambda, \Gamma)$ be the space of C^r cocycles $f : \Lambda \to \Gamma$ satisfying $f(\gamma x) = \gamma f(x)\gamma^{-1}$ for all $x \in \Lambda$, $\gamma \in \Gamma$. Note that for the next section only, C^1 will mean *Lipschitz*. We give $C^r_\Gamma(\Lambda, \Gamma)$ the usual C^r-topology.

If $f \in C^r_\Gamma(\Lambda, \Gamma)$ and we define $\Phi_f(x) = f(x)\Phi(x)$, then $\Phi_f \in \mathrm{Diff}^r_\Gamma(\Lambda)$ and Φ_f induces the map ϕ on Λ/Γ. Generally, suppose that Φ' is a Γ-equivariant diffeomorphism covering ϕ. If Φ' is sufficiently C^0-close to Φ, then it follows from Proposition 3.1.11 that $\Phi' = \Phi_{f'}$ for some near identity cocycle $f' \in C^r_\Gamma(\Lambda, \Gamma)$. We remark that this would not necessarily be true if the dimension of the Γ-orbits of points in M were allowed to vary (see [**21**]).

9.1. Stable ergodicity: Γ compact and connected

Recall that Λ_Π, the set of points in Λ with trivial isotropy, is an open and dense Γ-invariant subset of Λ. Since Λ_Π is contained in the principal isotropy stratum M_Π of M and the action of Γ on M_Π is assumed to be free, it follows that Λ_Π/Γ is a hyperbolic subset of the *manifold* M_Π/Γ. Strictly speaking we get a uniform hyperbolic structure on the complement in Λ/Γ of any neighborhood of the singular set Λ_S/Γ. Since there is a symbolic dynamics for Λ/Γ, it follows that there exist infinitely many transverse homoclinic points in Λ_Π/Γ which are homoclinic to periodic points in Λ_Π/Γ. Necessarily, the orbits of these homoclinic points are bounded away from the ϕ-invariant set Λ_S/Γ. Consequently, it follows from Smale's theorem [**38**, Theorem 6.5.5, Exercise 6.5.1], (see also [**29**, §2.1]) that Λ_Π/Γ contains a transitive subshift of finite type X with a hyperbolic structure. Since all Γ-principal bundles over a subshift of finite type are trivial, it follows that $\Phi : \Lambda|X \to \Lambda|X$ is a skew extension of X. The next Lemma is an immediate consequence of Propositions 6.1, 6.2 of [**29**].

LEMMA 9.1.1. *Let $r \in (0, s]$. Then there exists a C^r-open and dense subset $\mathcal{U}_r$ of $C^r_\Gamma(\Lambda, \Gamma)$ such that $\Phi_f : \Lambda|X \to \Lambda|X$ is transitive for all $f \in \mathcal{U}_r$. Moreover, if $r \geq 2$, then $\mathcal{U}_r$ is C^2-open in $C^r_\Gamma(\Lambda, \Gamma)$.*

Let $f \in C^r_\Gamma(\Lambda, \Gamma)$. Given $x \in \Lambda$, denote the closure of the Φ_f-orbit through x by $\overline{O}_f(x)$. It follows from the Γ-invariance of Φ_f that $\overline{O}_f(\gamma x) = \gamma \overline{O}_f(x)$ for all $\gamma \in \Gamma$. It is easy to see that if the ϕ-orbit of $p(x) \in \Lambda/\Gamma$ is dense, then $p(\overline{O}_f(x)) = \Lambda/\Gamma$. Note that this property does not depend on the choice of cocycle f.

DEFINITION 9.1.2 (cf [**29**, §6]). Let $x \in \Lambda$ and suppose that $p(x) \in \Lambda/\Gamma$ has a dense ϕ-orbit. We say that $\overline{O}_f(x)$ is a *maximal transitivity component* for Φ_f if whenever $\overline{O}_f(x') \cap \overline{O}_f(x) \neq \emptyset$, then $\overline{O}_f(x') \subset \overline{O}_f(x)$.

LEMMA 9.1.3 (cf [**29**, Lemma 6.5]). *Suppose that $\mathcal{P} = \{g\overline{O}_f(x) \mid g \in \Gamma\}$ defines a partition of Λ into maximal transitivity components. Then Φ_f is transitive if we can find a Φ_f-invariant and Γ-invariant closed subset Z of Λ such that $\Phi_f|Z$ is transitive.*

PROOF. Suppose $\Phi_f|Z$ is transitive. There exists $z \in Z$ such that the Φ_f-orbit of z is dense in Z. For some $h \in \Gamma$, $z \in h\overline{O}_f(x) = \overline{O}_f(hx)$. Hence, $\overline{O}_f(hx) \supset \overline{O}(z) = Z$. Therefore $\overline{O}_f(hx) \cap \overline{O}_f(gx) \neq \emptyset$, all $g \in \Gamma$ and so $\overline{O}_f(hx) = \Lambda$. □

THEOREM 9.1.4. *Let $r \in (0, s]$. Then there is a C^r-open and dense subset $\mathcal{U}_r$ of $C^r_\Gamma(\Lambda, \Gamma)$ such that (Φ_f, Λ, ν) is ergodic for all $f \in \mathcal{U}_r$. Moreover, if $r \geq 2$, then $\mathcal{U}_r$ is C^2-open in $C^r_\Gamma(\Lambda, \Gamma)$.*

PROOF. By Corollary 8.2.4, we have a partition of Λ into maximal transitivity components for all $f \in C^r_\Gamma(\Lambda, \Gamma)$. Now apply Lemmas 9.1.1, 9.1.3. □

REMARK 9.1.5. The proof of Corollary 8.2.4 depended on Proposition 8.2.1 which in turn depended on the strong stable and unstable foliations of Φ. Provided that Φ is C^1, these foliations exist. Moreover, the foliations continue to exist for Φ_f, $f \in C^r_\Gamma(\Lambda, \Gamma)$, $r \in (0, s]$ [**30**]. This suffices for the proof of Proposition 8.2.1. ◇

THEOREM 9.1.6. *Suppose that $(\phi, \Lambda/\Gamma, \mu)$ is topologically mixing. Let $r \in (0, s]$. For all $f \in \mathcal{U}_r$, (Φ_f, Λ, ν) is mixing and Bernoulli.*

Before proving this result we need some preliminary results (see Chapter 2, §2.1). Our approach is similar to that in [**31**].

Since Γ is connected, there is a finite covering homomorphism $\eta : \tilde{\Gamma} = \mathbb{K}^m \times S \to \Gamma$, where the group S is semisimple. Set $\tilde{S} = \eta(\{e_{\mathbb{K}^m}\} \times$

$S) \subset \Gamma$, $\tilde{A} = \eta(\mathbb{K}^m \times \{e_S\})$. Obviously $\tilde{S}, \tilde{A}$ are normal subgroups of Γ, $\tilde{S}$ is semisimple and $\tilde{A} \cong \mathbb{K}^m$.

Let $\pi : \Gamma \to \Gamma/\tilde{S}$ denote the quotient homomorphism. Set $\Gamma/\tilde{S} = A$ and note that $A \cong \mathbb{K}^m$. We remark that π is Γ-equivariant with respect to the given action of Γ on Γ ($g \mapsto \gamma g \gamma^{-1}$) and the trivial action on A.

LEMMA 9.1.7. *There is an open neighborhood $U = U^{-1}$ of the identity $e_A \in A$ and smooth Γ-equivariant local section $s : U \to \Gamma$ ($\pi \circ s = Id_U$).*

PROOF. This is a special case of a standard result in the theory of homogeneous spaces. In our situation, $\eta : \tilde{A} \to A$ is a finite covering map and the centralizer of $\tilde{A}$ in Γ is equal to $\tilde{S}$. Hence we can define s to be a local inverse to $\eta|\tilde{A}$ at the identity. □

LEMMA 9.1.8. *Let $r \geq 0$ and $f \in C^r_\Gamma(\Lambda, \Gamma)$. Set $\bar{f} = \pi \circ f \in C^r(\Lambda, A)$. There is a C^0 open neighborhood $\mathcal{W}_{\bar{f}}$ of $\bar{f}$ in $C^r(\Lambda, A)$ and continuous map $\chi : \mathcal{W}_{\bar{f}} \to C^r_\Gamma(\Lambda, \Gamma)$ such that $\chi(\bar{f}) = f$.*

PROOF. If f is constant, equal to the identity map of Γ, the result follows trivially from Lemma 9.1.7. For general f, we apply the special case to maps f' with $ff'^{-1} \in \mathcal{W}_{\bar{e}_A}$. □

LEMMA 9.1.9 ([**31**, Proposition 3.2.1]). *Suppose that $(\phi, \Lambda/\Gamma, \mu)$ is topologically mixing. Then stable ergodicity of (Φ_f, Λ, ν) implies stable mixing.*

PROOF. Suppose that Φ_f is stably ergodic but not mixing. It follows from Proposition 7.6.2 that there exists a nontrivial irreducible representation $\rho : \Gamma \to \mathbb{K}$, nonzero measurable $w : \Lambda \to \mathbb{C}$, and $c \in K$ such that

$$w \circ \Phi_f = cw, \quad w\gamma = \gamma w, \quad (\gamma \in \Gamma).$$

Since $(\phi, \Lambda/\Gamma, \mu)$ is ergodic, and w is Γ-equivariant, $w \neq 0$ on a full measure subset of Λ and so we may replace w by $w/|w|$ and assume that $w : \Lambda \to \mathbb{K}$.

Following our earlier notation, $\Gamma \supset \tilde{S}$, where $\tilde{S}$ is semisimple and $\Gamma/\tilde{S} = A \cong \mathbb{K}^m$. Since $\tilde{S}$ is semisimple, $\rho|\tilde{S}$ is constant, equal to the identity of $\mathbb{K}$. Hence ρ induces a homomorphism $\bar{\rho} : A \to \mathbb{K}$. Identifying A and $\mathbb{K}^m$, we set $\bar{f} = (\bar{f}_1, \dots, \bar{f}_m) = \pi \circ f \in C^r(\Lambda, \mathbb{K}^m)$. Writing $\Phi_f = f\Phi$, it follows that there exists $\ell = (\ell_1, \dots, \ell_m) \in \mathbb{Z}^m$, $\ell \neq 0$, such that

$$w \circ \Phi/w = c\Pi_{i=1}^m \bar{f}_i^{\ell_i}.$$

We are now in a position to use the argument of the proof of Proposition 3.2.1 [**31**]. We choose sequences $(b_n) \subset \mathbb{K}$, $(N(n)) \subset \mathbb{N}$, such that

$b_n \to 1$ and $(cb_n^{-\ell_1})^{N(n)} = 1$, all n. It follows that

$$\begin{aligned} w^{N(n)} \circ \Phi / w^{N(n)} &= \left(\frac{c}{b_n^{\ell_1}}\right)^{N(n)} b_n^{\ell_1 N(n)} \bar{f}_1^{\ell_1 N(n)} \Pi_{i=2}^m \bar{f}_i^{\ell_i N(n)}, \\ &= (b_n \bar{f}_1)^{\ell_1 N(n)} \Pi_{i=2}^m \bar{f}_i^{\ell_i N(n)}, \\ &= \Pi_{i=1}^m (\bar{f}_i')^{\ell_i}, \end{aligned}$$

where $\bar{f}_1' = b_n \bar{f}_1$, $\bar{f}_i' = \bar{f}_i$, $i > 1$. Applying Lemma 9.1.8, we see that for n sufficiently large

$$w^{N(n)} \circ \Phi_{f'} = w^{N(n)},$$

where $f' \in C_\Gamma^r(\Lambda, \Gamma)$ is the lift of $\bar{f}'$. Hence, by Proposition 7.6.1, $\Phi_{f'}$ cannot be ergodic. Letting $n \to \infty$, it follows from Lemma 9.1.8 that $\Phi_{f'} \to \Phi_f$, contradicting the stable ergodicity of Φ_f. □

Proof of Theorem 9.1.6 If Γ is semisimple, this result follows from Corollary 8.1.4. For general Γ, it follows from Lemma 9.1.9 that if $f \in \mathcal{U}_r$, then (Φ_f, Λ, ν) is mixing. The result now follows from Rudolph's theorem just as in the proof of Corollary 8.1.4. □

9.2. Stable ergodicity: Γ semisimple

THEOREM 9.2.1. *Suppose that Φ is C^s, $\infty \geq s \geq 1$. Let Γ be semisimple and $t \in (0, s]$. Then there exists a C^t-open and C^s-dense subset $\mathcal{U}_s$ of $C_\Gamma^s(\Lambda, \Gamma)$ such that (Φ_f, Λ, ν) is ergodic for all $f \in \mathcal{U}_s$. If the induced map ϕ on Λ/Γ is topologically mixing, then ergodic maps are Bernoulli.*

PROOF. We show that if Γ is semisimple, then there is a C^s-open and C^t-dense set of functions f such that (Φ_f, Λ, ν) is ergodic. Corollary 8.1.4 implies that if the induced map on Λ/Γ is topologically mixing then (Φ_f, Λ, ν) is Bernoulli.

It follows from Corollary 8.1.3 and Proposition 8.2.1(e), that it suffices to show that $Q_f(z) = \Lambda$ for a C^s-open, C^t-dense set of cocycles f, where $Q_f(z)$ is a transitivity component for the diffeomorphism Φ_f.

Fix $z \in \Lambda$. Since the periodic points of ϕ are dense in Λ/Γ by Theorem 5.7.5, we may choose $z_0 \in Q(z)$, such that $p(z_0)$ is periodic. We use the argument of Field and Parry [**31**, Theorem 4.3.1] based on homoclinic points (we might equally well have used the original argument of Brin [**12**, Page 10]).

It follows from Proposition 7.4.1 that we can choose points $z_1, z_2 \in \Lambda$, $p(z_0), p(z_1), p(z_2)$ distinct, and $\gamma, \eta \in \Gamma$ such that $z_1, z_2 \in Q(z_0)$ and

$$z_1 \in W_\varepsilon^{uu}(z_0), \quad \gamma z_0 \in W_\varepsilon^{ss}(z_1), \quad z_2 \in W_\varepsilon^{uu}(\gamma z_0), \quad \eta z_0 \in W_\varepsilon^{ss}(z_2).$$

The points γz_0, ηz_0 lie in $Q(z)$ since $Q(z)$ is a closed subset of Λ and is therefore partitioned by the leaves of the strong stable and unstable foliation. Hence, by Lemma 8.2.1(e), $Q(z)$ is invariant by the group $\langle \gamma, \eta \rangle$ topologically generated by γ, η.

If $\langle \gamma, \eta \rangle = \Gamma$, then $Q(z) = \Lambda$ and so Φ is ergodic. Let $\mathcal{U}_s$ denote the subspace of $C^s(\Lambda, \Gamma)$ consisting of cocycles f for which $\langle \gamma, \eta \rangle = \Gamma$ (fixed choice of $p(z_0)$, $p(z_1)$, $p(z_2)$ in Λ/Γ). It follows from the openness and density of topological generators for semisimple groups (Chapter 2, §2.1.1) that if $f' \in C^s(\Lambda, \Gamma)$ is sufficiently C^t-close to $f \in \mathcal{U}_s$, then $\Phi_{f'}$ will also be ergodic as the corresponding group elements $\gamma(f'), \eta(f')$ depend continuously on f' [**30**, Theorem 2.7.1] in the C^t-topology. In particular, $\mathcal{U}_s$ is C^t-open for all $t \in (0, s]$.

If $\langle \gamma, \eta \rangle$ is a proper subgroup of Γ, then we may make a C^s-small perturbation Φ_f of Φ supported on the Γ orbit of small $\bar{d}$ neighborhoods of $\{p(z_1), p(z_2)\}$ so that the corresponding pair of group elements $\gamma(f), \eta(f)$ generate Γ. Hence we have C^s-density. □

9.3. Stable ergodicity for attractors

It is rather easy to generalize Brin's original stability argument to the case when Λ is an attractor and Γ is an arbitrary compact connected Lie group. (Thanks are due to Keith Burns and Andrew Török for helpful conversations relating to this section.)

THEOREM 9.3.1. *Suppose that Φ is C^s, $s \geq 1$. Let Γ be a compact connected Lie group and $\Lambda \subset M$ be a (connected) attractor. For $r > 0$, there exists a C^0-open C^r dense subset $\mathcal{U}_r$ of $C^r_\Gamma(\Lambda, \Gamma)$ such that (Φ_f, Λ, ν) is ergodic for all $f \in \mathcal{U}_r$.*

PROOF. Since Λ is an attractor, Λ contains the unstable manifold (defined as a subset of M) of every point $x \in \Lambda$. In particular, $W^{uu}(x) \subset \Lambda$ is path connected for all $x \in \Lambda$.

Fix a point $z_0 \in \Lambda \setminus \Lambda_S$. Let R be a differentiable slice, radius $\varepsilon > 0$ through z_0 and set $U = \Gamma R$. Necessarily, U is disjoint from Λ_S. We assume in the sequel that ε is chosen so that $d(U, \Lambda_S) > 10\varepsilon$. Fix a Γ-equivariant diffeomorphism $\psi : U \to R \times \Gamma$ such that $\psi(z_0) = (p(z_0), e)$. Note that since we are assuming R is an ε-disk and so R is contractible. In the sequel, we identify R with $p(R) \subset M/\Gamma$ and regard U as a Γ-principal bundle over R. Our convention is that Γ acts on the *left* on the trivialization $R \times \Gamma$ (note that Brin [**12**] assumes Γ acts on the right). In addition, we have a right action of Γ on U, inherited from the right action on $R \times \Gamma$.

Let $\rho \in (0, \varepsilon)$, and set $U_\rho = \{z \in U \mid d(z, \Gamma z_0) < \rho\}$, $R_\rho = R \cap U_\rho$. For sufficiently small $\rho > 0$, we have

$$W^u_\varepsilon(\Gamma x) \cap W^s_\varepsilon(\Gamma y) = \Gamma z \subset U, \quad (x, y \in U_\rho).$$

In particular, we have a well-defined bracket operation $[\ ,\ ] : U_\rho^2 \to U$. Following [**12**, §2], let $\mathcal{L}^4$ denote the set of 5-tuples $(x_0, x_1, x_2, x_3, x_4) \in R_\rho^5$ such that

(a) $x_0 = x_4 = p(z_0)$.
(b) $x_{i+1} \in W^\sigma_\varepsilon(x_i)$, where $\sigma \in \{s, u\}$ and σ alternates round the loop.
(c) If $x_{i+1} \in W^u(x_i)$, then x_{i+1} is connected to x_i by a continuous path lying within $R_\rho \cap W^u(x_i)$.

Suppose that $L = (x_0, x_1, x_2, x_3, x_4) \in \mathcal{L}^4$. There is a natural lift of $(x_0, x_1, x_2, x_3, x_4)$ to $(z_0, z_1, z_2, z_3, z_4) \subset U_\rho$, where (a) $z_4 \in \Gamma z_0$, (b) $z_{i+1} \in W^{\sigma\sigma}_\varepsilon(z_i)$, where $\sigma \in \{s, u\}$ and σ alternates round the loop, (c) if $z_{i+1} \in W^{uu}(z_i)$, then z_{i+1} is connected to z_i by a continuous path lying within $U_\rho \cap W^{uu}(z_i)$.

Identifying Γz_0 with Γ via the map $\psi | \Gamma z_0$, the process of lifting defines a map $\pi : \mathcal{L}^4 \to \Gamma$ by $\pi(L) = \gamma$, where $\gamma z_0 = z_4$. If $L \in \mathcal{L}^4$, there is a continuous deformation of L to a degenerate path $L' \in \mathcal{L}^4$ with $\pi(L') = e$. Indeed, suppose that $L = (x_0, x_1, x_2, x_3, x_4) \in \mathcal{L}^4$ and, without loss of generality, assume that $x_1 \in W^u(x_0)$. Choose a continuous path $u_t \in W^u(x_0) \cap R_\rho$ such that $u_1 = x_1$, $u_0 = x_0$. Define $v_t = [u_t, x_3]$. Then v_t is a continuous path in $W^u(x_3) \cap R_\varepsilon$ and $v_1 = x_3$. In this way, we define a continuous family of paths $L_t \in \mathcal{L}^4$, with $L^0 = L$, $L' = (x_0, x_0, x_3, x_3, x_0)$. Obviously, $\pi(L') = e$.

Using the left Γ-action on U for composition of paths in $\mathcal{L}^4$, Let $H \subset \Gamma$ denote the group generated by $\{\pi(L) \mid L \in \mathcal{L}^4\}$. Every element of H may be obtained as the lift of a finite composition of elements of $\mathcal{L}^4$ (left Γ-action). In particular, H is path connected and so, by the theorem of Kuranishi-Yamabe [**77**], H has the structure of a Lie group. Specifically, there exists a Lie subalgebra $\mathfrak{h}$ of the Lie algebra of Γ such that $H = \exp(\mathfrak{h})$ (in general, H will not be a Lie *subgroup* of Γ).

The remainder of Brin's proof of genericity of stable ergodicity goes through without essential change and we refer the reader to [**12**, §2] for details. □

9.4. Stable ergodicity and SRB attractors

In this section we assume that $\Lambda \subset M$ is a transversally hyperbolic attractor for the C^s-diffeomorphism Φ, $s \geq 2$.

We suppose that we are given a Γ-regular Markov partition on Λ and associated subshift of finite type $\pi : \Sigma \to \Lambda/\Gamma$. Let λ denote Riemannian measure on M. There are several definitions of Sinai-Ruelle-Bowen (SRB) measures in the literature all of which are equivalent when Λ is uniformly hyperbolic. For our purposes, we use the following strong definition of SRB measure.

DEFINITION 9.4.1. Let ν be a Φ-invariant probability measure on Λ. We say that ν is an SRB measure if there exists an open neighborhood U of Λ and a λ-measure zero subset A of U such that for all continuous functions $f : U \to \mathbb{R}$ and all $x \in U \setminus A$,

$$\lim_{n\to\infty} \frac{1}{n} \sum_{i=0}^{n-1} f(\Phi^n(x)) = \int_\Lambda f d\nu.$$

We recall that $\mathrm{j}^u(x) = -\ln |\mathrm{Jac} D\Phi|E^u(x)|$, $x \in \Lambda$, is called the *Jacobian potential* and that j^u is Hölder continuous. Since j^u is Γ-invariant, j^u drops down to a Hölder continuous map on Λ/Γ which in turn lifts to a Hölder continuous potential on Σ. If we let $m^\star$ denote the corresponding unique equilibrium state on Σ, then the Haar lift of $\pi \cdot m^\star$ defines a Φ-invariant Borel probability measure $\nu^\star$ on Λ. The measure $\nu^\star$ is an equilibrium state of j^u.

In general, if ν is a Φ-invariant Borel probability measure on Λ we say that ν has *absolutely continuous conditional measures on unstable manifolds* if the conditional measures on strong unstable manifolds $W^{uu}(x)$ are absolutely continuous with respect to Lebesgue measure (on $W^{uu}(x)$).

It follows from the theorem of Ledrappier and Young [**44**, Theorem A] that a Φ-invariant Borel probability measure ν on Λ is an equilibrium state of j^u if and only if ν has absolutely conditional measures on the unstable manifolds of Λ.

PROPOSITION 9.4.2. *If $\nu^\star$ is an ergodic measure on Λ, then $\nu^\star$ is SRB.*

PROOF. Since $\nu^\star$ is an equilibrium state of j^u it follows from the theorem of Ledrappier and Young that $\nu^\star$ has absolutely continuous conditional measures on the unstable manifolds of Λ. Extending by Haar measure to the center unstable manifolds $W^{cu}(x)$, it follows that $\nu^\star$ has absolutely conditional measures on the center unstable manifolds. On the other hand, it follows from a theorem of Pesin [**64**], [**62**, Theorem 7.1] that the strong stable manifolds of points $x \in \Lambda$ are absolutely continuous. It is now routine and easy to show that if $\nu^\star$ is ergodic it is SRB. $\square$

For the remainder of this section we assume that Γ is compact and connected.

THEOREM 9.4.3. *Let* Φ *be* C^s, $s \geq 2$ *and* Λ *be a transversally hyperbolic attractor for* Φ.

(a) *If* $r \in [2, s]$, *there is a* C^2*-open,* C^r*-dense subset* $\mathcal{U}_r$ *of* $C^r(\Lambda, \Gamma)$ *such that if* $f \in \mathcal{U}_r$ *then* $(\Phi_f, \Lambda, \nu^\star)$ *is SRB.*

(b) *If* $(\Phi, \Lambda, \nu^\star)$ *is SRB and* V *is an open neighborhood of* Λ *in* M *then there exists a* C^2*-open neighborhood* $\mathcal{U}$ *of* Φ *in* $\mathit{Diff}_\Gamma(M)$ *such that if* $\Phi' \in U$, *then* Φ' *has a transversally hyperbolic SRB attractor* $\Lambda' \subset V$.

PROOF. The proof of part (a) of the theorem proceeds as in the proof of Part (a) of the Theorem 9.2.1 except that we have to use the version of Livšic regularity that holds for general equilibrium states of Hölder continuous potentials (see Remarks 8.1.2(1)).

Part (b) follows from general stability results on partially hyperbolic sets [**37**, §7] together with the arguments of the proof of Theorem 9.2.1. We omit the straightforward and routine arguments. □

APPENDIX A

On the absolute continuity of ν

Granted the existence of Γ-regular Markov partitions on Λ, the proof of absolute continuity of the stable and unstable foliations is a relatively straightforward variation of the argument used by Ruelle & Sullivan to prove the corresponding result for hyperbolic basic sets [**69**, Theorem 1, page 324]. We include some of the details for completeness and because we wish to emphasize the absolute continuity property that we use. While we give the result only for the measure of maximal entropy, our proof extends with minor modifications (the multiplying factors λ are no longer constant) to any Gibbs measure defined by a Hölder continuous potential.

Let m be the Parry measure on Ω and $\mu = \pi m$ be the induced ϕ-invariant measure on Λ/Γ of maximal entropy. Given $\varepsilon > 0$, $\alpha \in \Lambda/\Gamma$, we let $W_\varepsilon^{s/u}(\alpha)$ denote the local stable and unstable sets through α. Thus, if $x \in \alpha$, $W_\varepsilon^s(\alpha) = pW_\varepsilon^{ss}(x) = pW_\varepsilon^s(\Gamma x)$.

Following 7.4, we fix a Γ-regular Markov partition $\mathcal{R}$ of Λ. We suppose that for some $\varepsilon > 0$ $\mathcal{R}$ is ε-admissible and that $\text{mesh}(\mathcal{R}) < \varepsilon/3$. In what follows, the local stable and unstable sets will always be assumed of 'diameter' ε. Let $\tilde{\mathcal{R}}$ denote the Markov partition induced by $\mathcal{R}$ on Λ/Γ. Set $\Sigma^1 = W_\varepsilon^u(\Lambda_S) \cup W_\varepsilon^s(\Lambda_S)$ and $\tilde{\Sigma}^1 = p\Sigma^1$. We define $\Sigma = \cup_{n\in\mathbb{Z}} f^n(\Sigma^1)$, $\tilde{\Sigma} = p\Sigma$. Since $\nu(\Sigma^1) = 0$, $\nu(\Sigma), \mu(\tilde{\Sigma}) = 0$.

It follows from Lemma 6.5.6 (3) that we have a local product structure $[\![\ ,\ ]\!] : \tilde{R}^\sigma \times \tilde{R}^\sigma \to \tilde{R}$ characterized by $[\![\alpha, \beta]\!] = W^s(\alpha, \tilde{R}) \cap W^u(\beta, \tilde{R})$. It follows from Lemma 7.4.2 that this local product structure yields holonomy maps $h^\rho_{\alpha,\beta} : W^\rho(\alpha, \tilde{R}) \to W^\rho(\beta, \tilde{R})$, $\rho \in \{s, u\}$, whenever $\alpha, \beta \in \tilde{R}^\sigma$.

LEMMA A.1. *We may choose $\mathcal{R}$ and $\varepsilon > 0$ and a zero measure subset $\nabla \subset \Lambda/\Gamma$ so that there exists $\eta > 0$ such that if $\alpha, \beta \in \Lambda/\Gamma \setminus \nabla$ and $\bar{d}(\alpha, \beta) < \eta$, then there are well-defined μ-measurable holonomy maps*

$$h^u_{\alpha,\beta} : W^u_{\varepsilon/2}(\alpha) \to W^u_\varepsilon(\beta),\quad h^s_{\alpha,\beta} : W^s_{\varepsilon/2}(\alpha) \to W^s_\varepsilon(\beta).$$

In case $\alpha, \beta \in \tilde{R}^\sigma$, $R \in \mathcal{R}$, then $h^{u/s}_{\alpha,\beta}$ coincide with the holonomy maps defined previously on $W^{u/s}(\alpha, \tilde{R})$.

PROOF. Given $\varepsilon > 0$, we may choose $\eta > 0$ so that if $x, y \in \Lambda$ and $\bar{d}(x, y) < \eta$, then $[\![a, y]\!]$ is defined for all $a \in W^{ss}_{\varepsilon/2}(x)$ and $[\![a, y]\!] \in W^u_\varepsilon(y)$. However, as there may be multiple intersections between $W^{ss}_{\varepsilon/2}(x)$ and $W^u_\varepsilon(\Gamma y)$, this construction does not induce well defined holonomy maps on Λ/Γ unless we require $x, y \in \tilde{R}^\sigma$ and restrict to $a \in W^{uu}_\varepsilon(x) \cap R$. Suppose then that $\alpha, \beta \in \Lambda/\Gamma$, $\bar{d}(\alpha, \beta) < \eta$ and $\zeta \in W^u_{\varepsilon/2}(\alpha)$. Suppose there exist (a smallest) $N \geq 0$, $S \in \mathcal{R}$, $\tau \in W^u_\varepsilon(\beta)$ such that $\tau \in W^s_\varepsilon(\zeta)$ and $f^N(\zeta), f^N(\tau) \in \tilde{S}^\sigma$. Then we may pull back the holonomy map $h^u_{f^N(\zeta), f^N(\tau)}$ by f^{-N} to give a holonomy map defined on $f^{-N}(W^u_{\varepsilon/2}(\eta, \tilde{S}))$. Since the Markov partition is admissible, the holonomy map we obtain is uniquely defined. We need to show that the measure of the set ∇^u of α, β where this process fails is zero. Clearly, if $\alpha, \beta \notin \tilde{\Sigma}$, then the construction works provided only that there exists an $N > 0$ and $S \in \mathcal{R}$ such that $f^N(\zeta), f^N(\tau) \in \tilde{S}$. The only way this can fail is if in the limit $f^N(\zeta), f^N(\tau)$ converge to the union of the boundaries of the rectangles in $\tilde{\mathcal{R}}$. Since $\bar{d}(f^N(\zeta), f^N(\tau))$ converges zero exponentially fast and the measure of the boundary set is zero, we can use the argument of [**69**, Theorem 1, p324] together with $\mu(\tilde{\Sigma}) = 0$ to deduce that $\mu(\nabla^u) = 0$. The argument for the stable holonomy maps is the same and we conclude by setting $\nabla = \nabla^u \cup \nabla^s$. □

If we set $U = \{\alpha, \beta \in \Lambda/\Gamma \mid \bar{d}(\alpha, \beta) < \eta\}$, then it follows from Lemma A.1 that we have a measurable bracket operation $[\ ,\ ] : U \to \Lambda/\Gamma$ defined μ ae by

$$[\alpha, \beta] = h^u_{\alpha,\beta}(\alpha).$$

Given $\kappa > 0$, $\alpha \in \Lambda/\Gamma$, let $B_\kappa(\alpha) \subset \Lambda/\Gamma$ denote the open $\bar{d}$-disk of radius $\kappa > 0$, center α.

THEOREM A.2. *For μ-almost all $\alpha \in \Lambda/\Gamma$, there exist positive measures μ^s_α on $W^s_\varepsilon(\alpha)$, and μ^u_α on $W^u_\varepsilon(\alpha)$ such that*

(a) *$supp(\mu^u_\alpha) = W^u_\varepsilon(\alpha)$, $supp(\mu^s_\alpha) = W^s_\varepsilon(\alpha)$.*
(b) *$\mu | B_\eta(\alpha) = [\ ,\ ](\mu^u_\alpha \times \mu^s_\alpha) | B_\eta(\alpha)$.*

Furthermore, the holonomy maps $h^{u/s}_{\alpha,\beta}$ preserve $\mu^{u/s}_\alpha$-measure and, in particular, map sets of measure zero to sets of measure zero.

PROOF. Let A be the 01-matrix indexed by $\tilde{\mathcal{R}} \times \tilde{\mathcal{R}}$ with elements $a_{\tilde{S}_i, \tilde{S}_j}$. We let $\tilde{S}_i$ denote a variable member of the partition and use the notation $\tilde{R}_i$ when we are considering a specific member. Thus general elements of Ω will be written $(\tilde{S}_i)$. As in Ruelle and Sullivan [**69**], we avoid the use of double subscripts, such as $\tilde{R}_{i_k}$, when it is clear from context what is meant.

By the Perron-Frobenius theorem, A and its adjoint $A^\star$ have a dominant eigenvalue $\lambda > 0$ and corresponding eigenvectors v, w, with positive components v_i, w_i, such that $< v, w >= 1$. The Parry measure m is defined on cylinder sets by

$$m\{(\tilde{S}_i) \in \Omega \mid \tilde{S}_k = \tilde{R}_k, ..., \tilde{S}_l = \tilde{R}_l\} \quad = \quad \lambda^{-(l-k)} w_k a_{\tilde{R}_k, \tilde{R}_{k+1}} ... a_{\tilde{R}_{l-1}, \tilde{R}_l} v_l,$$

where $k \leq l$.

Let $\mathbb{Z}^+ = \{n \in \mathbb{Z} \mid n \geq 0\}$, $\mathbb{Z}^- = \{n \in \mathbb{Z} \mid n \leq 0\}$, $\Omega^+ = \{(\tilde{S}_i) \mid i \geq 0\}$ and $\Omega^- = \{(\tilde{S}_i) \mid i \leq 0\}$. Given $\mathbf{R} = (\tilde{R}_i) \in \Omega$ and $0 \leq k$, we define $\Omega^+_{\mathbf{R},k} = \{(\tilde{S}_i) \in \Omega \mid \tilde{S}_0 = \tilde{R}_0, ..., \tilde{S}_k = \tilde{R}_k\}$, $\Omega^-_{\mathbf{R},k} = \{(\tilde{S}_i) \in \Omega \mid \tilde{S}_{-k} = \tilde{R}_{-k}, ..., \tilde{S}_0 = \tilde{R}_0\}$. We define measures $m^\pm$ on $\Omega^\pm$ such that for $0 \leq k$ and $\mathbf{R} \in \Omega$,

$$\begin{aligned} m^+(\Omega^+_{\mathbf{R},k}) &= \lambda^{-k} a_{\tilde{R}_0, \tilde{R}_1} ... a_{\tilde{R}_{k-1}, \tilde{R}_k} v_k, \\ m^-(\Omega^-_{\mathbf{R},k}) &= \lambda^{-k} w_k a_{\tilde{R}_{-k}, \tilde{R}_{-k+1}} ... a_{\tilde{R}_{-1}, \tilde{R}_0}. \end{aligned}$$

Following [**69**], we show that $m^\pm$ induce measures on the (local) stable and unstable foliations of Λ. Thus, suppose $\alpha = \pi(\mathbf{R}) \in \Lambda/\Gamma$, where $\mathbf{R} = (\tilde{R}_i) \in \Omega$. We define $\pi^+_{\alpha,R_0} : \Omega^+_{\mathbf{R},0} \to \tilde{R}_0 \subset \Lambda/\Gamma$ by

$$\pi^+_{\alpha,R_0}(\tilde{S}_i) = \pi(..., \tilde{R}_{-2}, \tilde{R}_{-1}, \tilde{R}_0, \tilde{S}_1, \tilde{S}_2, ...).$$

The image of π^+_{α,R_0} is equal to $W^u_\varepsilon(\alpha, \tilde{R}_0)$. The image of $m^+|\Omega^+_{\mathbf{R},0}$ under π^+_{α,R_0} is a positive measure μ^u_{α,R_0} on $W^u_\varepsilon(\alpha, \tilde{R}_0)$. Since $\phi\pi^+_{\alpha,R_0}(S_i) = \pi^+_{\alpha^\star(S_1),S_1}(\sigma(S_i))$, for any $\alpha^\star(S_1) \in S_1 \cap \phi(W^u_\varepsilon(\alpha, R_0)$, the image of $m^+|\Omega^+_{\mathbf{R},1}$ by σ is given by $\lambda^{-1} m^+|\Omega_{\sigma\mathbf{R},0}$. Hence

$$\phi\mu^u_{\alpha,R_0} = \lambda^{-1} \sum_{R_1 \in \mathcal{R}} a_{R_0,R_1} \mu^u_{\alpha^\star(R_1),R_1}. \tag{A.1}$$

Next we must show that for almost all $\alpha \in \Lambda/\Gamma$, the measures μ^u_α, μ^s_α patch together on $W^{u/s}_\varepsilon(\alpha)$ to give measures invariant by the holonomy maps $h^{u/s}_{\alpha,\beta}$. But this follows from (A.1) using the same argument we used for the proof of Lemma A.1, which in turn was an extension of the original argument of Ruelle and Sullivan. The remaining statements of the theorem follow immediately from our construction. □

It follows from Theorem A.2 that for ν almost all $x \in \Lambda$ we may define the measures ν^u_x, ν^s_x on $W^u_\varepsilon(\Gamma x)$, $W^s_\varepsilon(\Gamma x)$ to be the Haar lifts of μ^u_{px}, μ^y_{px} respectively. Let ν^{uu}_x denote the conditional measure of ν^u_x on the strongly unstable fiber $W^{uu}_\varepsilon(x)$. Let [,] denote the ν-measurable lift of the bracket [,] defined on Λ/Γ. As an immediate consequence of our definitions and Theorem A.2 we obtain the following absolute continuity property needed for our proof of Theorem 8.1.1.

COROLLARY A.0.4. *For ν almost all $x \in \Lambda$, $[\ ,\ ](\nu_x^{uu} \times \nu_x^s)|B_\eta(x) = \nu|B_\eta(x)$.*

Bibliography

[1] R Adler, B Kitchens and B H Marcus. 'Finite group actions on shifts of finite type', *Erg. Thy. & Dynam. Sys.* **5** (1985), 1–25.

[2] R Adler, B Kitchens and M Shub. 'Stably ergodic skew products', *Discrete and Continuous Dynamical Systems* **2** (1996), 349–350.

[3] J F Alves, C Bonatti and M Viana. 'SRB measures for partially hyperbolic systems whose central direction is mostly expanding', *Invent. Math* **140** (2000), 351–398.

[4] E Bierstone. 'General position of equivariant maps', *Trans. Amer. Math. Soc.*, **234** (1977), 447–466.

[5] E Bierstone. 'Generic equivariant maps', *Real and Complex Singularities, Oslo 1976*, Proc. Nordic Summer School/NAVF Sympos. Math. (Sijthoff and Noordhoff International Publ.) Leyden (1977), 127–161.

[6] C Bonatti and M Viana. 'SRB measures for partially hyperbolic systems whose central direction is mostly contracting', *Israel Journal of Math.* **115** (2000), 157–194.

[7] R Bowen. 'Topological entropy and Axiom A', in *Global Analysis*, 23–41. Providence, AMS, 1970 (*proc. of Symp. on Pure Math. XIV.*

[8] R Bowen. 'Markov partitions for Axiom A diffeomorphisms', *Amer. J. Math.,* **92** (1970), 725–747.

[9] R Bowen. *Equilibrium States and the Ergodic Theory of Anosov Diffeomorphisms.* Springer Lect. Notes in Math. **470**, 1975.

[10] G E Bredon. *Introduction to compact transformation groups* (Pure and Applied Mathematics, **46**, Academic Press, New York and London, 1972.)

[11] M I Brin. 'Topology of group extensions of Anosov systems', *Mathematical Notes of the Acad. of Sciences of the USSR*, **18**(3) (1975), 858–864.

[12] M I Brin. 'Topological transitivity of one class of dynamic systems and flows of frames on manifolds of negative curvature', *Funkts. Anal. Prilozh*, **9**, No. 1, (1975), 9–19.

[13] M I Brin and J B Pesin. 'Partially hyperbolic dynamical systems', *Math. USSR Izvestia*, **8**, (1974), 177–218.

[14] T Bröcker and T Tom Dieck. *Representations of Compact Lie Groups* (Springer, New York, 1985).

[15] K Burns, C Pugh, M Shub and A Wilkinson. 'Recent results about stable ergodicity', *Proc. Symposia AMS*, **69** (2001), 327–366.

[16] K Burns and A Wilkinson. 'Stable ergodicity and skew products', *Ann. Sci. École Norm. Sup.* **32** (1999), 859–889.

[17] K Falconer. *Fractal Geometry- Mathematical Foundations and Applications* (Wiley, 1990).

[18] U-R Fiebig. 'Periodic points and finite group actions on shifts of finite types', *Erg. Thy. & Dynam. Sys.* **13** (1993), 485–514.

[19] M. J. Field. 'Transversality in G-manifolds', *Trans. Amer. Math. Soc.*, **231** (1977), 429–450.

[20] M J Field. 'Equivariant Dynamical Systems', *Trans. Amer. Math. Soc.*, **259** (1980),185–205.

[21] M J Field. 'On the structure of a class of equivariant maps', *Bull. Austral. Math. Soc.*, **26**(2) (1982), 161–180.

[22] M J Field. Isotopy and stability of equivariant diffeomorphisms, *Proc. London Math. Soc.*, **46**(3) (1983), 487–516.

[23] M J Field. 'Equivariant diffeomorphisms hyperbolic transverse to a G-action', *J. London Math. Soc.*, **27**(2) (1983), 563–576.

[24] M J Field. 'Equivariant Bifurcation Theory and Symmetry Breaking', *J. Dynamics and Diff. Eqns.*, **1**(4) (1989), 369–421.

[25] M J Field. 'Local structure of equivariant dynamics', in *Singularity Theory and its Applications, II*, eds. M. Roberts and I. Stewart, Springer Lecture Notes in Math., **1463** (1991), 168–195.

[26] M J Field. 'Symmetry breaking for compact Lie groups', *Mem. Amer. Math. Soc.* **574**, 1996.

[27] M J Field. 'Generating sets for compact semisimple Lie groups', *Proc. Amer. Math. Soc.* **127** (1999), 3361–3365.

[28] M J Field, I Melbourne and M Nicol. 'Symmetric Attractors for Diffeomorphisms and Flows', *Proc. London Math. Soc.* **72**(3) (1996), 657–696.

[29] M J Field, I Melbourne and A Török. 'Stable ergodicity for smooth compact Lie group extensions of hyperbolic basic sets', preprint 2003.

[30] M Field and V Niţică. 'Stable topological transitivity of skew and principal extensions', *Nonlinearity* **14** (2001), 1055–1070.

[31] M J Field and W Parry. 'Stable ergodicity of skew extensions by compact Lie groups', *Topology*, **38**(1) (1999), 167–187.

[32] W Fulton and J Harris. *Representation Theory*, Grad. Texts in Mathematics, Springer-Verlag, 1991.

[33] M Golubitsky, D G Schaeffer and I N Stewart. *Singularities and Groups in Bifurcation Theory, Vol. II*, (Appl. Math. Sci. **69**, Springer-Verlag, New York, 1988).

[34] J Guckenheimer and P Holmes. Structurally stable heteroclinic cycles, *Math. Proc. Camb. Phil. Soc.* **103** (1988), 189–192.

[35] M Grayson, C Pugh and M Shub. 'Stably ergodic diffeomorphisms', *Annal of Math.* **140** (1994), 295–329.

[36] M Hirsch and C Pugh. 'Stable manifolds and hyperbolic sets', *Proc. AMS Symp. in Pure Math.*, **14** (1970), 133–163.

[37] M W Hirsch, C C Pugh and M Shub. *Invariant manifolds*, (SLNM, Vol. 583. Springer-Verlag: New York, Heidelberg, Berlin, 1977)

[38] A Katok and B Hasselblatt. *Introduction to the Modern Theory of Dynamical Systems* (Encyclopedia of Mathematics and its Applications, **54**, Cambridge University Press, 1995.)

[39] K Kawakubo. *The theory of Transformation Groups* (Oxford University press, 1991).

[40] H B Keynes and D Newton. 'Ergodic measures for nonabelian compact group extensions', *Compositio Mathematica*,**32**, (1976), 53-70.

[41] M Krupa. 'Bifurcations of relative equilibria', *SIAM J. MATH. ANAL.*, **21**(6) (1990), 1453–1486.

[42] M Kuranishi. 'Two element generations on semisimple Lie groups', *Kodai. math. Sem. report*,(1949), 9-10.

[43] M G Nerurkar. 'Ergodic continuous skew-product actions of amenable groups', *Pac. J. Math*, **119**, (1985), 343-364.

[44] F Ledrappier and L S Young. 'The metric entropy of diffeomorphisms Part I: Characterization of measures satisfying Pesin's entropy formula', *Ann. of Math* **122** (1985), 509–539.

[45] A N Livšic. 'Cohomology of Dynamical Systems', *Mathematics of the USSR Izvestija*, **6**(6), (1972), 1278–1301.

[46] A N Livšic. 'Homology properties of Y-Systems', *Math. Notes*, **10**, (1971), 758–763.

[47] R Mañé. *Ergodic Theory and Differentiable Dynamics* (Springer-Verlag, Berlin, Heidelberg, New York, 1987).

[48] D Mongomery and L Zippin. *Topological Transformation Groups* (Interscience tracts in pure and applied Mathematics, **1**, Interscience, New Yorkm 1955).

[49] G D Mostow. 'Equivariant embeddings in euclidean space', *Ann. Math.* **65** (1957), 432–446.

[50] S E Newhouse. 'Lectures on Dynamical Systems', in *Dynamical Systems*, Progress in Mathematics **8**, Birkhäuser, 1980.

[51] M Nicol. 'Symmetries of the asymptotic dynamics of random compositions of equivariant maps', *Nonlinearity*, **9**, (1996), 225-235.

[52] M Nicol, M Pollicott. 'Measurable cocycle rigidity for some non-compact groups'. *Bull. LMS* **31**, (1999) 592–600.

[53] M Nicol, M Pollicott. 'Livsic's Theorems for Semisimple Lie Groups'. *Erg. Th. Dyn. Syst.* **21** (2001) 1501–1509.

[54] R S Palais. 'Imbedding of compact differentiable transformation groups in orthogonal representations', *J. Math. Mech.* **6** (1957), 673–678.

[55] R S Palais. 'On the existence of slices for actions of non-compact Lie groups', *Ann. Math.* **73** (1961), 295–323.

[56] J Palis. 'A global view of dynamics and a conjecture on the denseness of finitude of attractors', *Astérisque* **261** (2000), 335–347.

[57] W Parry. 'Skew-products of shifts with a compact Lie group', *J. London Math. Soc.*, **56**(2) (1997), 400–404.

[58] W Parry and M Pollicott. *Zeta functions and the periodic orbit structure of hyperbolic dynamics*, Astérisque **187-188**, Société Mathématique de France, 1990.

[59] W Parry and M Pollicott. 'The Livsic cocyle equation for compact Lie group extensions of hyperbolic systems', *J. London Math. Soc.* (2) **56** (1997), 405–416.

[60] W Parry and M Pollicott. 'Stability of mixing for toral extensions of hyperbolic systems', Tr. Mat Inst. Steklova **216** (1997), Din. Sist. i Smezhnye Vopr., 354–363.

[61] Ya B Pesin and Ya G Sinai. 'Gibbs measures for partially hyperbolic attractors', *Erg. Thy. & Dynam. Sys.* **2** (1982), 417–438.

[62] M Pollicott. *Lectures on ergodic theory and Pesin theory on compact manifolds* (LMS Lecture Note Series, **180**, Cambridge University Press, 1993.)

[63] M Pollicott and C Walkden. 'Livšic theorems for connected lie groups', *Trans. AMS.* (Volume 353) (7) (2001), 2879–2895.

[64] C Pugh and M Shub. 'Ergodic attractors', *Trans. Amer. Math. Soc.* **312** (1) (1989), 1–54.

[65] C Pugh and M Shub. 'Stably ergodic dynamical systems and partial hyperbolicity', *J. of Complexity* **13** (1997), 125–179.

[66] C Robinson. *Dynamical Systems*, CRC Press, Boca Raton, Ann Arbor, London, Tokyo, 1995.

[67] W Rudin. *Real and Complex Analysis.* (McGraw-Hill, Third Edition, 1987).

[68] D Rudolph. 'Classifying the isometric extensions of Bernoulli shifts', *J. d'Analyse Math.* **34** (1978), 36–60.

[69] D Ruelle and D Sullivan. 'Currents, flows and diffeomorphisms', *Topology*, **14**, (1975), 319-327

[70] G S Schwarz. 'Smooth functions invariant under the action of a compact Lie group', *Topology* **14** (1975), 63–68.

[71] G W Schwarz. 'Lifting smooth isotopies of orbit spaces', *Publ. I.H.E.S.*, **51** (1980), 37–135.

[72] M Shub and A Wilkinson. 'Pathological foliations and removable zero exponents', *Invent. Math.* **139** (2000), 495–508.

[73] P Walters. *An Introduction to Ergodic Theory.* (Springer Verlag, 1982).

[74] A Wilkinson. 'Stable ergodicity of the time-one map of a geodesic flow', *Erg. Th. Dyn. Syst.* **18**(6) (1998), 1545–1588.

[75] R F Williams. 'One-dimensional non-wandering sets', *Topology*, **6** (1967), 473–487.

[76] R F Williams. 'Classification of subshifts of finite type', *Annals of Math.*, **98** (1973), 120–153, and **99** (1974), 380–381

[77] H Yamabe. 'On an arcwise connected subgroup of a Lie group', *Osaka Math J.* **2** (1950), 13–14.

Editorial Information

To be published in the *Memoirs*, a paper must be correct, new, nontrivial, and significant. Further, it must be well written and of interest to a substantial number of mathematicians. Piecemeal results, such as an inconclusive step toward an unproved major theorem or a minor variation on a known result, are in general not acceptable for publication. Papers appearing in *Memoirs* are generally longer than those appearing in *Transactions*, which shares the same editorial committee.

As of February 1, 2004, the backlog for this journal was approximately 4 volumes. This estimate is the result of dividing the number of manuscripts for this journal in the Providence office that have not yet gone to the printer on the above date by the average number of monographs per volume over the previous twelve months, reduced by the number of volumes published in four months (the time necessary for preparing a volume for the printer). (There are 6 volumes per year, each containing at least 4 numbers.)

A Consent to Publish and Copyright Agreement is required before a paper will be published in the *Memoirs*. After a paper is accepted for publication, the Providence office will send a Consent to Publish and Copyright Agreement to all authors of the paper. By submitting a paper to the *Memoirs*, authors certify that the results have not been submitted to nor are they under consideration for publication by another journal, conference proceedings, or similar publication.

Information for Authors

Memoirs are printed from camera copy fully prepared by the author. This means that the finished book will look exactly like the copy submitted.

The paper must contain a *descriptive title* and an *abstract* that summarizes the article in language suitable for workers in the general field (algebra, analysis, etc.). The *descriptive title* should be short, but informative; useless or vague phrases such as "some remarks about" or "concerning" should be avoided. The *abstract* should be at least one complete sentence, and at most 300 words. Included with the footnotes to the paper should be the 2000 *Mathematics Subject Classification* representing the primary and secondary subjects of the article. The classifications are accessible from `www.ams.org/msc/`. The list of classifications is also available in print starting with the 1999 annual index of *Mathematical Reviews*. The Mathematics Subject Classification footnote may be followed by a list of *key words and phrases* describing the subject matter of the article and taken from it. Journal abbreviations used in bibliographies are listed in the latest *Mathematical Reviews* annual index. The series abbreviations are also accessible from `www.ams.org/publications/`. To help in preparing and verifying references, the AMS offers MR Lookup, a Reference Tool for Linking, at `www.ams.org/mrlookup/`. When the manuscript is submitted, authors should supply the editor with electronic addresses if available. These will be printed after the postal address at the end of the article.

Electronically prepared manuscripts. The AMS encourages electronically prepared manuscripts, with a strong preference for $\mathcal{A}_{\mathcal{M}}\mathcal{S}$-LaTeX. To this end, the Society has prepared $\mathcal{A}_{\mathcal{M}}\mathcal{S}$-LaTeX author packages for each AMS publication. Author packages include instructions for preparing electronic manuscripts, the *AMS Author Handbook*, samples, and a style file that generates the particular design specifications of that publication series. Though $\mathcal{A}_{\mathcal{M}}\mathcal{S}$-LaTeX is the highly preferred format of TeX, author packages are also available in $\mathcal{A}_{\mathcal{M}}\mathcal{S}$-TeX.

Authors may retrieve an author package from e-MATH starting from `www.ams.org/tex/` or via FTP to `ftp.ams.org` (login as `anonymous`, enter username as password, and type `cd pub/author-info`). The *AMS Author Handbook* and the *Instruction Manual* are available in PDF format following the author packages link from `www.ams.org/tex/`. The author package can be obtained free of charge by sending email to `pub@ams.org` (Internet) or from the Publication Division, American Mathematical Society, 201 Charles St., Providence, RI 02904, USA. When requesting an author package, please specify AMS-LaTeX or AMS-TeX, Macintosh or IBM (3.5) format, and the publication in which your paper will appear. Please be sure to include your complete mailing address.

Sending electronic files. After acceptance, the source file(s) should be sent to the Providence office (this includes any TeX source file, any graphics files, and the DVI or PostScript file).

Before sending the source file, be sure you have proofread your paper carefully. The files you send must be the EXACT files used to generate the proof copy that was accepted for publication. For all publications, authors are required to send a printed copy of their paper, which exactly matches the copy approved for publication, along with any graphics that will appear in the paper.

TeX files may be submitted by email, FTP, or on diskette. The DVI file(s) and PostScript files should be submitted only by FTP or on diskette unless they are encoded properly to submit through email. (DVI files are binary and PostScript files tend to be very large.)

Electronically prepared manuscripts can be sent via email to `pub-submit@ams.org` (Internet). The subject line of the message should include the publication code to identify it as a Memoir. TeX source files, DVI files, and PostScript files can be transferred over the Internet by FTP to the Internet node `e-math.ams.org` (130.44.1.100).

Electronic graphics. Comprehensive instructions on preparing graphics are available at `www.ams.org/jourhtml/graphics.html`. A few of the major requirements are given here.

Submit files for graphics as EPS (Encapsulated PostScript) files. This includes graphics originated via a graphics application as well as scanned photographs or other computer-generated images. If this is not possible, TIFF files are acceptable as long as they can be opened in Adobe Photoshop or Illustrator. No matter what method was used to produce the graphic, it is necessary to provide a paper copy to the AMS.

Authors using graphics packages for the creation of electronic art should also avoid the use of any lines thinner than 0.5 points in width. Many graphics packages allow the user to specify a "hairline" for a very thin line. Hairlines often look acceptable when proofed on a typical laser printer. However, when produced on a high-resolution laser imagesetter, hairlines become nearly invisible and will be lost entirely in the final printing process.

Screens should be set to values between 15% and 85%. Screens which fall outside of this range are too light or too dark to print correctly. Variations of screens within a graphic should be no less than 10%.

Inquiries. Any inquiries concerning a paper that has been accepted for publication should be sent directly to the Electronic Prepress Department, American Mathematical Society, 201 Charles St., Providence, RI 02904, USA.

Editors

This journal is designed particularly for long research papers, normally at least 80 pages in length, and groups of cognate papers in pure and applied mathematics. Papers intended for publication in the *Memoirs* should be addressed to one of the following editors. In principle the Memoirs welcomes electronic submissions, and some of the editors, those whose names appear below with an asterisk (*), have indicated that they prefer them. However, editors reserve the right to request hard copies after papers have been submitted electronically. Authors are advised to make preliminary email inquiries to editors about whether they are likely to be able to handle submissions in a particular electronic form.

Titles in This Series

803 **Michael Field and Matthew Nicol,** Ergodic theory of equivariant diffeomorphisms: Markov partitions and stable ergodicity, 2004

802 **Martin W. Liebeck and Gary M. Seitz,** The maximal subgroups of positive dimension in exceptional algebraic groups, 2004

801 **Fabio Ancona and Andrea Marson,** Well-posedness for general 2×2 systems of conservation laws, 2004

800 **V. Poénaru and C. Tanasi,** Equivariant, almost-arborescent representations of open simply-connected 3-manifolds; A finiteness result, 2004

799 **Barry Mazur and Karl Rubin,** Kolyvagin systems, 2004

798 **Benoît Mselati,** Classification and probabilistic representation of the positive solutions of a semilinear elliptic equation, 2004

797 **Ola Bratteli, Palle E. T. Jorgensen, and Vasyl' Ostrovs'kyĭ,** Representation theory and numerical AF-invariants, 2004

796 **Marc A. Rieffel,** Gromov-Hausdorff distance for quantum metric spaces/Matrix algebras converge to the sphere for quantum Gromov-Hausdorff distance, 2004

795 **Adam Nyman,** Points on quantum projectivizations, 2004

794 **Kevin K. Ferland and L. Gaunce Lewis, Jr.,** The $RO(G)$-graded equivariant ordinary homology of G-cell complexes with even-dimensional cells for $G = \mathbb{Z}/p$, 2004

793 **Jindřich Zapletal,** Descriptive set theory and definable forcing, 2004

792 **Inmaculada Baldomá and Ernest Fontich,** Exponentially small splitting of invariant manifolds of parabolic points, 2004

791 **Eva A. Gallardo-Gutiérrez and Alfonso Montes-Rodríguez,** The role of the spectrum in the cyclic behavior of composition operators, 2004

790 **Thierry Lévy,** Yang-Mills measure on compact surfaces, 2003

789 **Helge Glöckner,** Positive definite functions on infinite-dimensional convex cones, 2003

788 **Robert Denk, Matthias Hieber, and Jan Prüss,** $\mathcal{R}$-boundedness, Fourier multipliers and problems of elliptic and parabolic type, 2003

787 **Michael Cwikel, Per G. Nilsson, and Gideon Schechtman,** Interpolation of weighted Banach lattices/A characterization of relatively decomposable Banach lattices, 2003

786 **Arnd Scheel,** Radially symmetric patterns of reaction-diffusion systems, 2003

785 **R. R. Bruner and J. P. C. Greenlees,** The connective K-theory of finite groups, 2003

784 **Desmond Sheiham,** Invariants of boundary link cobordism, 2003

783 **Ethan Akin, Mike Hurley, and Judy A. Kennedy,** Dynamics of topologically generic homeomorphisms, 2003

782 **Masaaki Furusawa and Joseph A. Shalika,** On central critical values of the degree four L-functions for GSp(4): The Fundamental Lemma, 2003

781 **Marcin Bownik,** Anisotropic Hardy spaces and wavelets, 2003

780 **S. Marmi and D. Sauzin,** Quasianalytic monogenic solutions of a cohomological equation, 2003

779 **Hansjörg Geiges,** h-principles and flexibility in geometry, 2003

778 **David B. Massey,** Numerical control over complex analytic singularities, 2003

777 **Robert Lauter,** Pseudodifferential analysis on conformally compact spaces, 2003